TEN GREAT IDEAS ABOUT CHANCE

TEN GREAT IDEAS ABOUT CHANCE

PERSI DIACONIS
BRIAN SKYRMS

PRINCETON UNIVERSITY PRESS
Princeton and Oxford

Published by Princeton University Press,
41 William Street, Princeton, New Jersey 08540

In the United Kingdom: Princeton University Press,
6 Oxford Street, Woodstock, Oxfordshire OX20 1TR

press.princeton.edu

Jacket image courtesy of Shutterstock

ISBN 978-0-691-17416-7

Library of Congress Control Number 2017943311

British Library Cataloging-in-Publication Data is available

This book has been composed in Sabon Next LT Pro text with
Akzidenz Grotesk BQ Display

Printed on acid-free paper. ∞

Printed in the United States of America

1 3 5 7 9 10 8 6 4 2

WE DEDICATE THIS BOOK

TO THE MEMORY OF

Richard "Diamond Jim" Jeffrey,
good friend and true philosopher.

CONTENTS

PREFACE

This book grew out of a course that we taught together at Stanford for about 10 years. This was a large, mixed course. There were undergraduates and graduates. There were participants from philosophy, statistics, and a number of other disciplines across the academic spectrum. As the course evolved over time, we came to believe that the story we are telling would be of interest to a larger audience. Our course had as a prerequisite exposure to one course in probability or statistics. Our book retains this level. But for the reader who may have had such a course a long time ago, we have included an appendix designed as a probability refresher.

This is a history book, a probability book, and a philosophy book. We give the history of what we see as great ideas in the development of probability, but we also pursue the philosophical import of these ideas. One reader of an earlier version of this manuscript complained that at the end of the book, he still did not know our philosophical views about chance. We were, perhaps, too evenhanded. This problem has now been fixed. You will see that we are thorough Bayesians, followers of Bayes, Laplace, Ramsey, and deFinetti. Bayesianism is sometimes thought to be opposed to frequencies. We insist that our view does not deny the importance of frequencies or the usefulness of talking about objective chances. Rather, it unifies these considerations within the framework of rational degrees of belief.

At the beginning of this book we are thinking along with the pioneers, and the tools involved are simple. By the end, we are up to the present, and some technicalities have to be at arm's length. We try to ease the flow of exposition by putting some details in appendices,

which you can consult as you wish. We also try to provide ample resources for the reader who finds something interesting enough to dig
deeper. There is a select annotated bibliography at the end. There are
more detailed references in the footnotes.

Persi Diaconis
Brian Skyrms

ACKNOWLEDGMENTS

Many people have helped us in the creation of this book. First of all, we want to thank our ten generations of students for their constructive feedback. Many friends read all or parts of the manuscript, made helpful suggestions, and corrected errors. We especially want to thank Susan Holmes, Christian Robert, Steve Evans, Jim Pitman, Steve Stigler, David Mermin, Simon Huttegger, Jeff Barrett, and Sandy Zabell. Our editor, Vickie Kearn, and the team at Princeton helped turn a preliminary typescript into the finished book that you see.

TEN GREAT IDEAS ABOUT CHANCE

Gerolamo Cardano

CHAPTER 1

MEASUREMENT

One way to understand the roots of a subject is to examine how its originators thought about it. Some basic philosophical issues are already evident at the very beginning. The first great idea is simply that chance can be measured. It emerged during the sixteenth and seventeenth centuries, and it is something of a mystery why it took so long. The Greeks had a goddess of chance, Tyche. Democritus and his followers postulated a physical chance affecting all the atoms that made up the universe. This is the "swerve" of atoms in Lucretius' *De Rerum Natura*. Games of chance, using knucklebones or dice, were known to Egyptians and Babylonians and were popular in Rome. Soldiers cast lots for Christ's cloak. Greek Skeptics of the later Academy postulated probability (eikos) as the guide to life.[1] Nevertheless, it appears that there was no quantitative theory of chance in these times.[2]

Figure 1.1. Determination of the lawful rood

How do you measure anything?[3] Consider length. You find a standard of equal length, apply it repeatedly, and count. The standard might be your foot, as you pace off a distance. Different feet may not lead to the same result. One refinement, proposed in 1522 for determining a lawful rood (rod), was to line up the feet of 16 people as they emerged from church, as shown in figure 1.1.[4] As the illustration shows, the various folks have very different foot lengths, but an implicit averaging effect was accepted by a group—even though the explicit notion of an average seems to not have existed at the time.

It is worth mentioning a certain philosophical objection at this point. There is a kind of circularity involved in the procedure. We are defining length, but we are already assuming that our standard remains the same length as we step off the distance.

No sensible person would let this objection stop her from stepping off distance. That is how we start. Eventually we refine our notion of length. Your foot may change length; so may the rod; so may the standard meter stick, at a fine-enough precision. Using physics, we refine the measurement of length.[5] So the circularity is real, but it indicates a path for refinement rather than a fatal objection.*

So it is with chance. To measure probability, we first find—or make—equally probable cases. Then we count them. The probability of an event A, denoted by $P(A)$, is then

$$P(A) = \frac{\text{no. of cases in which } A \text{ occurs}}{\text{total no. of cases}}.$$

Note that it follows that

1. Probability is never negative,
2. If A occurs in all cases, $P(A) = 1$,
3. If A and B never occur in the same case,

$$P(A \text{ or } B) = P(A) + P(B).$$

In particular, the probability of an event not occurring is 1 less the probability of its occurring:

$$P(\text{not } A) = 1 - P(A).$$

It is surprising how much can be done by ingenious application of this simple idea. Consider the birthday problem. What is the probability that at least two people in a room share the same birthday, neglecting leap years, assuming birthdates are equiprobable and birthdays of individuals in the room are independent (no twins)? If you have not seen it before, the results are a bit surprising.

The probability of a shared birthday in the group is 1 minus the probability that they are all different. The probability that the second person has a different birthday from the first is $\left(\frac{364}{365}\right)$. If they are different, the probability that the third is different from them is $\left(\frac{363}{365}\right)$, and so on, for all in the room. So the probability of a shared birthday among N people is

$$1 - \left(\frac{364}{365} \cdot \frac{363}{365} \cdot \ldots \cdot \frac{365 - N + 1}{365} \right).$$

---

*What are the paths open for refinement of the notion of equiprobable? They will unfold as we move through the book.

If you are interested in an even-money bet, this formula can be used to find a value of N such that the product is close to $\frac{1}{2}$. If there are 23 people in the room, the probability of a shared birthday is slightly greater than $\frac{1}{2}$. If there are 50 people, it is close to 97%.

There are many variations on the birthday problem. These are used for thinking about surprising coincidences. For instance, it is overwhelmingly likely that there are two people in the United States who share a birthday, whose fathers share the same birthday, whose fathers' fathers share this birthday, and so on, four generations back. Useful approximations for working with these variations may be found in an appendix to this chapter. These approximations are, in turn, used to prove de Finetti's representation theorem in an appendix at the end of this book. The point for now is that the basic "equally likely cases" structure has real breadth and strength.

BEGINNINGS

Nothing provides us better candidates for equiprobable cases than vigorous throws of symmetric dice or draws from a well-shuffled deck of cards. This is where the measurement of probability began. We cannot say who was there first, but the idea was clearly there in the sixteenth-century work on gambling by the algebraist, physician, and astrologer Gerolamo Cardano.[6] Cardano, who sometimes made a living at gambling, was quite sensitive to the equiprobability assumption. He knew about shaved dice and dirty deals: ". . . the die may be dishonest either because it has been rounded off, or because it is too narrow (a fault which is easily visible), or because it has been extended in one direction by pressure on the opposite faces. . . . There are even worse ways of being cheated at cards."[7]

In the early seventeenth century Galileo composed a short note on dice to answer a question posed to him (by his patron, the Grand Duke of Tuscany). The Duke believed that counting possible cases seemed to give the wrong answer. Three dice are thrown. Counting combinations of numbers, 10 and 11 can be made in 6 ways, as can 9 and 12. ". . . yet it is known that long observation has made dice-players consider 10 and 11 to be more advantageous than 9 and 12."* How can this be?

Galileo replies that his patron is counting the wrong thing. He counts three 3s as one possibility for making a 9 and two 3s and a 4 as one possibility for making a 10. Galileo points out the latter covers three possibilities, depending on which die exhibits the 4:

<4, 3, 3>, <3, 4, 3>, <3, 3, 4>.

For the former, there is only <3, 3, 3>. Galileo has a complete grasp of permutations and combinations and does not seem to regard it as anything new.

In constructing equiprobable cases, both Galileo and Cardano appear to make implicit use of *independence*. They suppose that for each die, all 6 faces are equally probable and that for throws of 3 dice, all 216 possible outcomes are also equally probable. When we treated the birthday problem earlier, we assumed that different people had independent chances for their birthdays.

With this basic machinery well understood, Pascal and Fermat in their famous correspondence attacked more subtle problems with a different conceptual flavor.

PASCAL AND FERMAT (1654)

The first substantial work in the mathematics of probability appears to be the correspondence between Pascal and Fermat, which began in 1654. We include a discussion for three reasons: (1) It *is* the first; (2) it shows how seemingly complex problems can be reduced to straightforward calculations with equally likely cases; and (3) it introduces the crucial notion of expectation—a mainstay of the subject.

---

*One strange aspect of the statement of the problem is the comment about long observation. The observation would have had to be long indeed. From Galileo's calculations, the chance of a 9 is $\frac{25}{216}$, about 0.116; the chance of a 10 is $\frac{27}{216}$, about 0.125. The difference between these is 0.009, or about $\frac{1}{100}$. As an exercise, you could calculate how many observations would be required.

Pascal and Fermat addressed problems with a different conceptual flavor from those solved by Cardano and Galileo, defining fairness and focusing on expectation.

There are two problems given to Pascal by his sometime gambling friend, the Chevalier de Méré. Pascal communicated these, together with his thoughts on them, to Fermat, with whom he had a connection through the academy of Father Mersenne. This was the *Académie Parisienne* that Mersenne formed in 1635 where the work of leading mathematicians, scientists, and philosophers—including Galileo, Descartes, and Leibniz—was shared.

The *problem of dice*: A player has undertaken to throw a 6 in 8 throws of a die. The stakes have been settled, and the 3 throws have been taken without obtaining a 6. What proportion of the stake would be fair to give the player to forego his fourth throw (just the fourth).*

The *problem of points*: Two players of equal skill[†] are playing a series of games. The one to win a round gets a point. They have agreed that the first to reach a certain number of points wins the game and collects the stakes. A certain number of rounds have been played, and the game is interrupted. What is a fair division of the stakes?

Both of these problems are stated in terms of fairness. But what is fairness in the theory of probability? We will see that Pascal and Fermat implicitly employ the concept of *expectation* to answer that question.

The expected value of a gamble that pays off $V(x)$ in outcome x is the probability weighted average:

$$\text{expectation } (V) = V(x_1)\, p(x_1) + V(x_2)\, p(x_2) + \cdots .$$

A transaction that leaves the players' expected values unchanged is assumed to be fair. For example, consider flipping a fair coin. If it comes up heads you win 1; if tails you lose 1. Then the expected value is $(+1)(\frac{1}{2}) + (-1)(\frac{1}{2}) = 0$.

Let's apply this idea to the *problem of dice*. The stakes, s, are still on the table. If the player does not forego his fourth throw, he has 5 throws remaining. His expectation is

---

* You might try to think about this directly before you proceed. Suppose that before starting, bets are laid so that there is $10 on the table that you will get a 6 in eight throws. Would you take $5 to forego just trial 4? Would this be fair?
† We may as well think of them as flipping a fair coin.

$$\tfrac{1}{6}s \qquad\qquad + \qquad\qquad \tfrac{5}{6}\left(1-\left(\tfrac{5}{6}\right)^4\right)s$$

(win in fourth throw)   (lose on fourth but win on 1
of the remaining 4 throws).*

Suppose that that player foregoes his fourth throw for $\tfrac{1}{6}$ of the stakes, as Fermat suggests in the correspondence.[8] Then his expectation is

$$\tfrac{1}{6}s \qquad\qquad + \qquad\qquad \left(1-\left(\tfrac{5}{6}\right)^4\right)\left(\tfrac{5}{6}\right)s$$

(the amount   (probability of winning in
he is paid to forego   the remaining 4 throws times
that throw)   the diminished stakes).

These are the same, so $\tfrac{1}{6}$ of the stakes is the fair price for foregoing the throw.[9]

The *problem of points* is also an expected-value problem. It had baffled many previous thinkers. In 1494, Fra Luca Pacioli considered a problem of points where the play is complete with 6 points; one player has won 5 and the other, 3. Pacioli—perhaps under the influence of Aristotle's proportional theory of justice—argues that the fair division is in proportion to the rounds already won, 5 to 3. About 50 years later, Tartaglia objected that, according to this rule, if the game were stopped after 1 round, 1 player would be awarded the whole stake. This consequence looks worse and worse as the number of points necessary to win is increased. Tartaglia tried to modify Pacioli's rule to take this into account, but in the end he doubted that a definitive answer was possible. The problem puzzled all who thought about it, including Cardano and the Chevalier de Méré.

Fermat had the key insight. Suppose that one player needs r points to win and the other, s. Then the game will surely be decided in $r+s-1$ rounds. It may be decided earlier, but there is no harm in considering all sequences of $r+s-1$ coin flips, since the outcome is well defined for each. This reduces the problem to one of equiprobable cases and we can calculate probabilities by counting.

So for Pacioli's problem, where player 1 has 5 points and player 2 has 3, since 6 points concludes the game, 3 more rounds will suffice. Of

---

* Probability of winning on one of the remaining 4 throws $= 1 - P(\text{lose on all 4}) = (1-(\tfrac{5}{6})^4)$.

the 8 equiprobable cases, player 2 will win the game only if he wins all 3 rounds. His expectation is $\frac{1}{8}$ of the stakes, while player 1 has an expectation of $\frac{7}{8}$ on the stakes. It is fair, then, to divide the stakes in this proportion.

Expectation, computed by counting equiprobable cases, solves the problem. But there may be a large number of equiprobable cases to count. Consider Tartaglia's example. Six points win, and one player has no points and the other, 1 point. Then play must be complete after 10 more rounds. It would be tedious to write out the 1024 possible outcomes. But Pascal had a better way of counting.

To count the cases in which the first player wins, one adds the number of cases in which she gets 6 wins in 10 trials [called 10 choose 6] + the number where she gets 7 wins in 10 trials [10 choose 7] + $\cdots$ + the number where she gets 10 wins in 10 trials [10 choose 10]. These numbers are conveniently to be found on the tenth row of Pascal's arithmetical triangle (or Tartaglia's triangle, or Omar Khayyam's triangle[10]), which we show in figure 1.2. The row tells us the number of ways we can choose from a group of 10 objects. Reading from the left, there are 1 way of choosing nothing, 10 ways of choosing 1 object, 45 ways of choosing 2 objects, 120 ways of choosing 3, and so on, to only 1 way of choosing 10.

We want the number of ways of getting 6 wins in 10 trials + the number of ways to get 7 wins in 10 trials + $\cdots$ + the number where she gets 10 wins in 10 trials. From row 10 we get

$$210 + 120 + 45 + 10 + 1 = 386$$

for a probability of winning of

$$\frac{386}{1024} \quad \text{(about 38\%)}.$$

Thus a fair division of the stakes gives player 1 (who had no points) $\frac{386}{1024}$ of the stakes and player 2 the rest.

After Pascal and Fermat, the basic elements of measuring probability by counting equiprobable cases, calculating by combinatorial principles, and using expected value are all on the table.

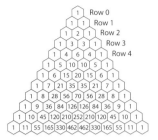

Figure 1.2. Pascal's triangle

HUYGENS (1657)

The ideas in the Pascal-Fermat correspondence were taken up and developed by the great Dutch scientist Christiaan Huygens[11] after he heard about the correspondence on a visit to Paris. He then worked them out by himself and wrote the first book on the subject in 1656. It was translated into English by John Arbuthnot in 1692 as *Of the Laws of Chance*.[12]

Huygens begins his book with a fundamental principle:

Postulat

As a Foundation to the following Proposition, I shall take Leave to lay down this Self-evident Truth: That any one Chance or Expectation to win any thing is worth just such a Sum, as wou'd procure in the same Chance and Expectation at a fair Lay. As for Example, if any one shou'd put 3 Shillings in one Hand, without letting me know which, and 7 in the other, and give me Choice of either of them; I say, it is the same thing as if he shou'd give me 5 Shillings; because with 5 Shillings I can, at a fair Lay, procure the same even Chance or Expectation to win 3 or 7 Shillings.

Huygens assumes that he could, in effect, flip a fair coin to choose which hand to pick.* Then $(\frac{1}{2})3 + (\frac{1}{2})7 = 5$. He then says that the value

---

*A point made much later by Howard Raiffa against the so-called Ellsberg paradox, which we will visit in our chapter on psychology of chance (chapter 3).

of the wager is the same as the value of 5 for sure. Thus he makes explicit (a special case of) the principle that is implicit in Pascal and Fermat: *expectation is the correct measure of value.*

He then goes on to justify this measure by a fairness argument. Suppose I bet 10 shillings with someone on the flip of a fair coin. This is fair by reasons of symmetry. Now suppose we modify this by an agreement that whoever wins shall give 3 to the loser. This preserves symmetry, so the modified arrangement is also fair. But now the loser nets 3 and the winner retains 7. Any such agreement preserves fairness, including where the winner gives the loser 5, and each has 5 for sure. Huygens then shows how the argument generalizes to arbitrary finite numbers of outcomes and arbitrary rational-valued probabilities of outcomes. It will be a recurring theme that an equality is justified by a symmetry.

NEWTONIAN CONSIDERATIONS

In the preface to the translation of Huygens, Arbuthnot, who was a follower of Newton,[13] makes the following noteworthy remark (L. Todhunter, *A History of the Mathematical Theory of Probability* (Cambridge: Macmillan, 1865); reprinted by Chelsea (New York, 1965), p. 51):

> It is impossible for a Die, with such determin'd force and direction, not to fall on such determin'd side, only I don't know the force and direction which makes it fall on such determin'd side, and therefore I call it Chance, which is nothing but the want of art.

Arbuthnot thus introduces the fundamental question of the proper conception of chance in a deterministic setting. His answer is that chance is an artifact of our ignorance.

Consider tossing a coin just once. The thumb hits the coin; the coin spins upward and is caught in the hand. It is clear that if the thumb hits the coin in the same place with the same force, the coin will land with the same side up. Coin tossing is physics, not random! To demonstrate this, we had the physics department build us a coin-tossing machine. The coin starts out on a spring, the spring is released, the coin spins upward and lands in a cup, as shown in figure 1.3. Because the forces are controlled, the coin always lands with the same side up.

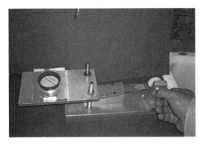

Figure 1.3. A deterministic coin-tossing machine

This is viscerally quite disturbing (even to the two of us). Magicians and crooked gamblers (including one of your authors) have the same ability.

How then is the probabilistic treatment of coin flips so widespread and so successful? The basic answer is due to Poincaré. If the coin is flipped vigorously, with sufficient vertical and angular velocity, there is sensitive dependence on initial conditions. Then a little uncertainty as to initial conditions is amplified to a large uncertainty about the outcome, where equiprobability of outcomes is not such a bad assumption. But the provisos are important. See appendix 2 for a little more on this. We will return to the question in more detail in our chapter on physical chance (chapter 9).

BERNOULLI 1713

In 1713 Jacob Bernoulli's *Ars Conjectandi*[14] was published, 8 years after his death. Bernoulli made explicit the practice of his predecessors. The first part is a reprint, with commentary, of Huygens. The probability of an event is now explicitly defined as the ratio of the number of (equiprobable) cases in which the event happens to the total number of (equiprobable) cases. The probability of being dealt a club from a deck of cards is $\frac{13}{52}$. He also defines the *conditional probability* of a second event (B) conditional on a first (A) as the ratio of the number of cases both happen to the number of cases the first happens:

$$\text{Probability } (B \text{ conditional on } A) = \frac{\text{no. of cases in which } A \text{ and } B \text{ occur}}{\text{no. of cases in which } A \text{ occurs}}.$$

The probability of being dealt a queen given that one is dealt a club is $\frac{1}{13}$.

On the basis of these definitions, he shows that the probabilities of mutually exclusive events add and that probabilities satisfy the multiplicative law, $P(A \text{ and } B) = P(A)P(B \text{ conditional on } A)$. These simple rules form the heart of all calculations of probability.

But Bernoulli's major contribution was to establish a rigorous connection between probability and frequency that had heretofore only been conjectured. He called this his golden theorem.

As an illustration he considers an urn containing 3000 white pebbles and 2000 black pebbles and postulates independent draws with replacement of the pebble drawn. He asks whether one can find a number of draws so that it becomes "morally certain" that the ratio of white pebbles to black ones becomes approximately 3:2. He then chooses a high probability as moral certainty and establishes a number of draws sufficient to provide a positive answer. Then he shows the weak law of large numbers:

> Given any interval around the probability (here $\frac{3}{5}$) as small as you please and any approximation to certainty, $1 - e$, as close as you please, there is a number of trials, N, such that in N trials the probability that the relative frequency of draws of white falls within the specified interval is at least $1 - e$.

This is a story to which we will return in our chapter on frequency (chapter 4).

SUMMING UP

Probability, like length, can be measured by dividing things into equally likely cases, counting the number of successful cases and dividing by the total number of cases. This definition satisfies the following:

1. Probability is a number between 0 and 1.
2. If A never occurs, $P(A) = 0$. If A occurs in all cases, $P(A) = 1$.
3. If A and B never occur in the same case, then $P(A \text{ or } B) = P(A) + P(B)$.

4. Conditional probability for B given A is defined by counting all the cases in which B and A occur together and dividing by the number of cases in which A occurs. Then, $P(A$ and $B) = P(A)P(B$ conditional on A). If A and B are independent, that is, if $P(B$ conditional on A) just equals $P(B)$, then $P(A$ and $B) = P(A)P(B)$.

Finding the cases and doing the counting leads to math problems such as the probability of winning a complicated wager or the birthday problem.

Expectation, weighting the costs and benefits of various outcomes by their chances, is useful for calculations *and* is a measure of fairness and value.

The law of large numbers, to which we will return in chapter 4 (and again in chapter 6), proves that chances can be approximated (with high probability) by frequencies in repeated independent trials.

APPENDICES

These three appendices give, respectively, a more detailed look at the correspondence between Pascal and Fermat, a development of the physics of coin tossing, and a more detailed analysis of the connection between the mathematics of probability and the real-world occurrence of chance events. (For those who might find it useful, there is a probability-refresher appendix at the end of this book.)

APPENDIX 1. PASCAL AND FERMAT

THE PROBLEM OF DICE

Pascal's first letter to Fermat is lost, but it must state the problem of the dice.

Fermat's reply points out that Pascal has made an error ("Pascal and Fermat on Probability," tr. by Vera Sanford in *A Sourcebook in Mathematics*, ed. David Eugene Smith (New York: McGraw Hill,

1929), 546–65. Dover reprint in 1969 available online at https://www
.york.ac.uk/depts/maths/histstat/pascal.pdf):

> If I undertake to make a point with a single die in eight throws,
> and if we agree after the money is put at stake, that I shall
> not cast the first throw, it is necessary by my theory that I
> take $\frac{1}{6}$ of the total sum to [be] impartial because of the aforesaid
> first throw.
>
> And if we agree after that, that I shall not play the second
> throw, I should, for my share, take the sixth of the remainder
> that is $\frac{5}{36}$ of the total.
>
> If, after that, we agree that I shall not play the third throw, I
> should to recoup myself, take $\frac{1}{6}$ of the remainder, which is $\frac{25}{216}$ of
> the total.
>
> And if subsequently, we agree again that I shall not cast the
> fourth throw, I should take $\frac{1}{6}$ of the remainder or $\frac{125}{1296}$ of the
> total, and I agree with you that that is the value of the fourth
> throw supposing that one has already made the preceding plays.
>
> But you proposed in the last example in your letter (I quote
> your very terms) that if I undertake to find the six in eight throws
> and if I have thrown three times without getting it, and if my
> opponent proposes that I should not play the fourth time, and
> if he wishes me to be justly treated, it is proper that I have $\frac{125}{1296}$
> of the entire sum of our wagers.
>
> This, however, is not true by my theory. For in this case, the
> three first throws having gained nothing for the player who holds
> the die, the total sum thus remaining at stake, he who holds the
> die and who agrees to not play his fourth throw should take $\frac{1}{6}$
> as his reward. And if he has played four throws without find-
> ing the desired point and if they agree that he shall not play the
> fifth time, he will, nevertheless, have $\frac{1}{6}$ of the total for his share.
> Since the whole sum stays in play it not only follows from the
> theory, but it is indeed common sense that each throw should be
> of equal value.

It is clear that the central issue here is that of *expected value*. The combination of foregoing a round and receiving a proportion of the stake
is fair if it leaves the expected value of the game unchanged.

Fermat sees clearly that the analysis is the same at any point in the game. Suppose that after the round in question, there will be $n+1$ rounds remaining; give the stakes at this point value 1. Then the value of taking the play is $\frac{1}{6}$ for winning now and $(\frac{5}{6})(1-(\frac{5}{6})^n)$ for failing on this throw but possibly eventually winning. The value of taking $\frac{1}{6}$ of the stakes and proceeding with the rest of the game for the diminished stakes is $\frac{1}{6}$ for the cash in hand plus $1-(\frac{5}{6})^n$, the probability of eventually, winning times $\frac{5}{6}$ of the the diminished stakes. Pascal immediately agrees with Fermat's analysis.

THE PROBLEM OF POINTS

There is another aspect of Pascal's discussion that is of interest. He starts with the example of a game where two players play for 3 points, where each has staked 32 pistoles ("Pascal and Fermat on Probability," tr. by Vera Sanford in *A Sourcebook in Mathematics*, ed. David Eugene Smith (New York: McGraw Hill, 1929), 546–65. Dover reprint in 1969 available online at https://www.york.ac.uk/depts/maths/histstat /pascal.pdf):

> Let us suppose that the first of them has two (points) and the other one. They now play one throw of which the chances are such that if the first wins, he will win the entire wager that is at stake, that is to say 64 pistoles. If the other wins, they will be two to two and in consequence, if they wish to separate, it follows that each will take back his wager that is to say 32 pistoles.
>
> Consider then, Monsieur, that if the first wins, 64 will belong to him. If he loses, 32 will belong to him. Then if they do not wish to play this point, and separate without doing it, the first should say "I am sure of 32 pistoles, for even a loss gives them to me. As for the 32 others, perhaps I will have them and perhaps you will have them, the risk is equal. Therefore let us divide the 32 pistoles in half, and give me the 32 of which I am certain besides." He will then have 48 pistoles and the other will have 16.

This is not just a calculation of expected value but also a justification of the *fairness* of using it, in terms that are hard for anyone to reject. What you have for sure is yours. For what is uncertain, equal

probabilities match equal division. It is a definitive answer to Fra Pacioli's line of thought.

Pascal goes on to show how this reasoning can be further iterated:

> Now let us suppose that the first has *two* points and the other
> *none*, and that they are beginning to play for a point. The
> chances are such that if the first wins, he will win all of the
> wager, 64 pistoles. If the other wins, behold they have come
> back to the preceding case in which the first has *two* points and
> the other *one*.
>
> But we have already shown that in this case 48 pistoles will
> belong to the one who has *two* points. Therefore if they do not
> wish to play this point, he should say, "If I win, I shall gain all,
> that is 64. If I lose, 48 will legitimately belong to me. Therefore
> give me the 48 that are certain to be mine, even if I lose, and let
> us divide the other 16 in half because there is as much chance
> that you will gain them as that I will." Thus he will have 48 and 8,
> which is 56 pistoles.
>
> Let us now suppose that the first has but *one* point and the
> other *none*. You see, Monsieur, that if they begin a new throw,
> the chances are such that if the first wins, he will have *two* points
> to *none*, and dividing by the preceding case, 56 will belong to
> him. If he loses, they will [be] point for point, and 32 pistoles will
> belong to him. He should therefore say, "If you do not wish to
> play, give me the 32 pistoles of which I am certain, and let us di-
> vide the rest of the 56 in half. From 56 take 32, and 24 remains.
> Then divide 24 in half, you take 12 and I take 12 which with 32
> will make 44.

This gives us a recursive procedure for fair division. Pascal then proj-
ects to games with larger numbers of points, and comes to a general
solution of the problem.

APPENDIX 2. PHYSICS OF COIN TOSSING

Drawing balls from an urn, flipping coins, rolling dice, and shuffling
cards are basic probability models. How are they connected to their

parallels in the real world? Going further afield, these basic models are often used to calculate chances in much more complicated setups; Bernoulli considered the successive scores of two tennis players. Gilovitch, Tversky, and Valone[15] considered the successive hits and misses of basketball players. Shouldn't physics and psychology come into these analyses?

Each of the foregoing examples has its own literature. To give a flavor of this, we consider a single flip of a coin. Afterward, pointers to the analysis of other examples will be given.

Let's take a brief look at a simple version of the physics.[16] When the coin leaves the hand, it has an initial velocity upward v (feet/second) and a rate of spin ω (revolutions/second). If v and ω are known, Newton tells us how much time the coin will take before landing and thus heads or tails are determined. The phase space of a coin in this model is thus as shown in figure 1.4.

A single flip corresponds to a point in this plane. Consider the point in figure 1.4. The velocity is large (so the coin goes up rapidly), but the rate of spin is low. Thus the coin goes up like a pizza tossed in the air, hardly turning. Similarly, a point with v small and ω large may be turning like crazy but never goes high enough to turn over once. From these considerations, it follows that there is a region of initial conditions, close to the two axes, where the coin never turns.

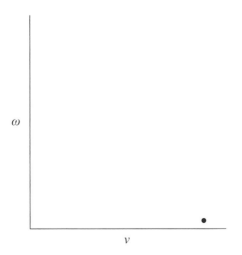

Figure 1.4. The $v\omega$-plane with a single flip

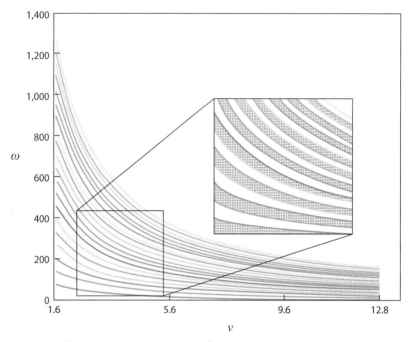

Figure 1.5. The hyperbolas separating heads from tails in part of phase space. Initial conditions leading to heads are hatched, tails are left white, and ω is measured in s^{-1}.

There is an adjoining region where the coin turns once, then a region for two turns, and so on. The full picture is shown in figure 1.5.

Inspection of the picture (and some easy mathematics) shows that regions far from 0 get closer together. So small changes in initial conditions make for the difference between heads or tails.

To go further, one must know the answer to the following question: When real people flip real coins, where are the points on the picture? We have carried out experiments and a normal flip takes about $\frac{1}{2}$ second and turns at about 40 revolutions/second. Look at figure 1.5. In the units of the picture, velocity is about $\frac{1}{5}$, very close to zero. The rate of spin, ω, is 40 units up, however, way off the picture. The math behind the picture says how close the regions are. This coupled with experimental work shows that coin tossing is fair to two decimal places but not to three.

The preceding analysis is in a simple model, which assumes that the coin flips about an axis through the coin. In fact, real coins are more complicated. They precess in amazing ways. A full analysis, with many details, caveats, and full references is in "Dynamical Bias in the Coin Toss,"[17] which concludes that vigorous tosses of ordinary coins are *slightly biased*. The chance of the coin landing the same way it started is about 0.51.

Where does all this analysis leave us? The standard model is a very good approximation. It would take about 250,000 flips to detect the difference between 0.50 and 0.51 (in the sense of giving second-digit accurately). We wish some of the other instances of the standard model were as solidly useful. Similar statements hold for Galileo's dice, but roulette or shuffling cards is another story![18]

If an honest analysis of a simple coin flip leads us into such complications, how much more would be required for an analysis of chances in games of skill or for the application of probability to medicine and law, as envisioned by Leibniz and Bernoulli? Bernoulli appreciated the point (Jacob Bernoulli, *The Art of Conjecturing*, tr. with an introduction and notes by Edith Dudley Sylla (Baltimore: Johns Hopkins University Press, 2006), 327):

> But what mortal, I ask, may determine, for example, the number of diseases, as if they were just as many cases, which may invade at any age the innumerable parts of the human body and which imply our death? And who can determine how much more easily one disease may kill than another—the plague compared to dropsy, dropsy compared to fever? Who, then, can form conjectures on the future state of life and death on this basis? Likewise who will count the innumerable cases of the changes to which the air is subject every day and on this basis conjecture its future constitution after a month, not to say after a year?
>
> Again, who has a sufficient perspective on the nature of the human mind or on the wonderful structure of the body so that they would dare to determine the cases in which this or that player may win or lose in games that depend in whole or in part on the shrewdness or the agility of the players? In these and similar situations, since they may depend on causes that are entirely hidden and that would forever mock our diligence by an

innumerable variety of combinations, it would clearly be mad to want to learn anything in this way.[19]

Bernoulli thought he had an answer to these problems in his law of large numbers. We return to this issue in chapter 4. There, and in subsequent chapters, we will assess the adequacy of the answer and discuss the possible alternatives.

APPENDIX 3. COINCIDENCES AND THE BIRTHDAY PROBLEM

Coincidences occur to all of us. Should we be surprised, or worried? The simple birthday problem (and its variations) has emerged as a useful tool to enable standards of surprise. While most people *are* surprised when there is a birthday match within a group of 23 people, the easy calculation in the introduction to this chapter shows that it is not surprising at all. Let us abstract and extend this calculation.

Consider what we will call the "watch" problem. Old-fashioned watches—watches with second hands—are coming back into current fashion. We believe that the second hands are "random"—completely out of sync and equally likely to show anything from 1 to 60—independently from watch to watch. Consider a group of N people, each having a watch with a second hand. What is the probability that two or more of these match—say, right at this second?

This is the birthday problem with 60 categories. The original birthday problem has 365 categories. Abstracting, consider C categories (so $C = 60$ for watches but $C = 365$ for birthdays.) There are N people, each independently and uniformly distributed in $\{1, 2, 3, \ldots, C\}$. What is the chance that all these numbers are distinct? Of course, this depends on C and N; the chance is zero if $N = C + 1$.

Call this chance $P(C, N)$. By our earlier reasoning,

$$P(C, N) = 1 \cdot \left(1 - \frac{1}{C}\right) \cdot \left(1 - \frac{2}{C}\right) \cdots \left(1 - \frac{(N-1)}{C}\right).$$

This is a neat formula. You can use it with a pocket calculator to give an exact answer for any fixed C and N.

But this is not particularly useful for understanding. For later use, we can compute a simple approximation, which shows that when $N = 1.2\sqrt{C}$, the chance of success is close to $\frac{1}{2}$. For the watch problem, $1.2\sqrt{60} = 9.3$, so a match has at least even odds with 10 people. Intuitively, a match would seem a striking coincidence. (For the original birthday problem, $1.2\sqrt{365} = 22.9$.)

We state our approximation as a proposition.

Proposition: With N people and C possibilities, N and C large, the chance of no match is

$$P(C, N) \sim e^{-N(N-1)/2C}.$$

Proof: The argument uses simple properties of the logarithm: $\log(1 - x) \sim -x$ when x is small. Then

$$P(C, N) = \left(1 - \frac{1}{C}\right) \cdot \left(1 - \frac{2}{C}\right) \cdot \cdots \cdot \left(1 - \frac{(N-1)}{C}\right)$$
$$= e^{\log(1 - 1/C) + \log(1 - 2/C) + \cdots + \log(1 - (N-1)/C)}$$
$$\sim e^{-1/C - 2/C \cdots - (N-1)/C}$$
$$= e^{-N(N-1)/2C}.$$

The approximations are accurate provided that N and C are large, with $N^{2/3}/C$ small.

Diaconis and Mosteller[20] use the birthday problem more generally in studying coincidences. They use these ideas to study multiple coincidences. For instance, how large should N be to have approximately even odds of a triple birthday match? (Answer: about 81.)

As a counterpoint to a philosophy that tries to make much of coincidences, we have provided a simple chance model for comparison. It seems useful and believable for studying things like a birthday match in a classroom. But one might consider instead a group of people in a very fancy restaurant. Since people are often taken out to dinner on their birthdays, it is quite likely that there may be several matches on a given night. The assumptions of our chance model don't hold, so the conclusions aren't relevant. The caution applies to all the simple chance models of this section. For more, see Diaconis and Holmes (2002).[21]

Frank Plumpton Ramsey

CHAPTER 2

JUDGMENT

Our second great idea is that judgments can be measured and that coherent judgments are probabilities. (What exactly is meant by coherence will be made clear.) In the classic gambling games of chapter 1, our judgments were informed by symmetry. Symmetrical cases were judged equiprobable. In this chapter we will see how degrees of belief implicit in judgments about all sorts of cases can also be measured. When they are so measured, coherent judgments turn out to have the same mathematical structure as that discovered by Cardano and Galileo by counting equiprobable outcomes in gambling.

How can we measure the probability that there will be a financial crisis in the next year, that the patient will survive if given this treatment, or that the defendant is guilty? How can we measure the probability that this candidate will win the election, that there will be a depression, or that reckless politics will precipitate a war? Here we do not have a nice set of intuitively equiprobable cases that allowed us to calculate by counting with supposedly perfectly fair dice. But, in fact, the law, politics and medicine were areas where Leibniz and Bernoulli envisioned the most important application of the calculus of probabilities. These probabilities can be no better than our degrees of belief based on the best available evidence.[1] That does not mean that they cannot be measured.

Below, we will be talking about assessing probabilities by betting. There are a number of real world instances in which you can do just that. Perhaps the simplest are prediction markets. These are Web sites in which you can bet for or against well-specified events, things like who will win a football game, or a horse race, or the next election. Prediction markets are not a new invention; in the sixteenth century there were markets for betting on who would be elected Pope.[2] In a typical prediction market, contracts are scaled between 0 and 100. At any time you can see offers to buy and to sell, say at 56.8 and 57.2. If you wish to buy a contract, you can buy instantly at the posted price of 57.2 or post an offer to buy, say, at 57.0, and wait to see whether anyone is willing to sell at that price. If you buy a contract on Clinton winning at 57, that means that for $57, you get a contract that pays $100 if she wins. Of course, the prices fluctuate.

It is very natural to take the current market price as the market's probability. The expected value of a bet that pays off $100 if C; $0 otherwise is $57, if the probability of C is 0.57. If the market prices do not obey the mathematics of probability, then—as we shall see—the market can be arbitraged. We think that *your* prices (I'll buy in for a small amount if the price is below x and sell if the price is above y), are good indications of *your* probability.

There is a healthy emerging literature on prediction markets. Buying stocks, bonds, and insurance are closely related activities. In all of these, the principles we lay out next can be useful.[3]

We divide the body of this chapter in two—between a naive and a sophisticated approach. The initial treatment will assume, like the early gamblers, that for the problems at issue, money is the relevant measure of value. That allows a straightforward way to measure judgmental probability and to infer the mathematical structure of such judgments provided that they are coherent. The second half of the chapter lifts this assumption and gives a more general analysis. This completion of the great idea measures both probability and utility at the same time. The leading ideas of the theory were given in the 1920s by a young genius, Frank Plumpton Ramsey, and fully developed in the 1950s by Leonard Jimmie Savage.[4]

PART I: GAMBLING AND JUDGMENTAL PROBABILITIES

To measure judgmental probabilities, we invert the approach of Pascal and Fermat and follow Huygens. Instead of measuring probabilities to compute expected value, we use expected value to measure probabilities. We measure the expected value imputed to an event by measuring the price that an individual will pay for a wager on that event. Your judgmental probabilities are then the quantities which, when used in a weighted average, give that expected value.

In particular,

The probability of A is just the expected value of a wager that pays off 1 if A and 0 if not.

If you pay a price equal to $P(A)$ for such a wager, you believe that you have traded equals for equals. For a lesser price you would prefer to buy the wager; for a greater price you would prefer not to buy it. So the balance point, where you are indifferent between buying the wager or not, measures your judgmental probability for A.

It is farfetched idealization to assume that people can effortlessly and reliably make such fine discriminations. But taking the first steps of the approximation is perhaps all we need to do for many decisions. How far can we go? There is no clear answer. We proceed to explore the theory that results from the full idealization.

COHERENT JUDGMENTS

Do judgmental probabilities, in general, have the mathematical structure gotten in chapter 1 by counting? Bruno de Finetti showed that if an individual's betting behavior is *coherent*, her judgmental probability, so defined, does indeed have the mathematical structure of a probability. The basic argument can be given very simply.*

Here is our idealized model—not meant to be all of real life but nevertheless meant to be instructive. An individual acts like a bookie—or perhaps like a derivatives trader—and buys and sells bets. She judges a bet as *fair* if her expected value for it is zero, *favorable* if her expected value is positive and disadvantageous if her expected value is negative. She buys fair or favorable bets and sells fair or disadvantageous bets, doing business with all comers. A *Dutch book* can be made against her if there is some finite set of transactions acceptable to her such that she suffers a net loss in every possible situation. We will say that she is *coherent* if she is not susceptible to a Dutch book.

An Example: Suppose you ask me for my judgmental probability that Senator Foghorn will win a second term. After some thought, I say 0.6. Then you ask me for my probability that Bobbie Blowhard will be elected instead. I quickly say 0.1. Then I am asked for the probability that either Foghorn or Blowhard will win, and I say 0.9. If I stick to these judgmental probabilities I am incoherent. You can make a Dutch book against me by buying from me a bet that pays off 1 if Foghorn wins for 0.6 and a bet that pays off 1 if Blowhard wins for 0.1 and then turning around and selling me a bet that pays 1 if either wins for 0.9. You are covered no matter who wins and pocket the profit of 0.2.

If you are kind enough to point this out to me instead of exploiting my incoherence, I may well reconsider my probability judgments. We all make careless judgments that are full of incoherence. Sometimes it doesn't matter much. But what if the stakes were high enough to be important? To take an extreme case, suppose you are a hedge-fund manager and there are other hedge-fund managers in the market. If you were made aware of your incoherence, wouldn't you tend to do a little rethinking?

Aiming for coherence has its roots in a desire for consistency. It applies to logic as well. One of the wisest men we know put it this way: "We all believe inconsistent things. The purpose of rational

---

*At this point the argument proceeds as if we could just use money as a measure of value. This assumption will be lifted in the second half of this chapter.

discussion aims at this: If someone says 'You accept *A* and *B*, but by a chain of reasoning, each step of which you accept, it can be shown that *A* implies not *B*,' you would think that something is wrong and want to correct it."

It is similar with judgments of uncertainty. Of course, there is no bookie, and no one is betting. Still coherence, like consistency, seems like a worthwhile standard.

De Finetti showed that coherence is equivalent to one's judgments having the mathematical structure of probability.

COHERENT JUDGMENTS ARE PROBABILITIES

To say one's judgments have this mathematical structure is just to say that they behave as proportions. They are proportions of partial belief. Proportions have a minimum of 0. A tautology, which is true throughout the whole space of possibilities, has proportion equal to 1. Proportions of a combination of mutually exclusive parts—of jointly inconsistent propositions—add. If there are 20% red beans and 35% white beans in a bag then there are 55% beans that are red or white. As we will see presently, that is all we need for the mathematical structure of probability applied to a finite space of possibilities.[5]

    I. Coherence implies probability.

       1. Minimum of zero.

           Suppose you give some proposition, *p*, a probability less than 0. Then you will give a bet where you *lose 1 if p, nothing otherwise* a positive expected value.

           You will then suffer a net loss no matter what happens. If *p* doesn't happen, you gain nothing from the bet and lose what you paid for it. If *p* does happen, you have a double loss. You lose what you paid for the bet and you lose the bet as well.

       2. Tautology gets probability of 1.

           Suppose that you give a tautological proposition, one that is true in any case, a probability different than 1. It is either greater than or less than 1. If it is greater than 1, you would pay more than 1 for a bet that pays off *1 if p, nothing otherwise*. When bets are settled you would win only 1, for a net loss. If you were to give the tautology probability less than 1, you would sell a bet that pays off *1 if p, nothing otherwise* for less than 1. When bets are settled, your buyer would collect 1 and you would be left with a net loss.

3. Mutually exclusive parts add.

Now consider additivity. Suppose p, q, are mutually exclusive. (Their conjunction is inconsistent.) But notice that an arrangement that pays off *1 if p or q; nothing otherwise* can be made in two different ways. It can be made directly, as a bet on p or q, or it can be achieved indirectly by simultaneous bets on p and on q.

The cost of the indirect bets held together is $P(p) + P(q)$ and coherence requires that the cost of the direct bet should be the same: $P(p \text{ or } q) = P(p) + P(q)$. If we should have inequality in either direction, a Dutch book could obviously be made by buying cheap and selling dear. Within the assumptions of the model, coherence is essentially a matter of consistent evaluations of a betting arrangement that can be implemented in more than one way.

II. Probability implies coherence.

If judgments are probabilities, then expected values of bets add.

$$E(b_1 + b_2 + \cdots) = E(b_1) + E(b_2) + \cdots$$

But a Dutch book requires that $E(b_1 + b_2 + \cdots)$ be negative because of the sure loss, while $E(b_1)$, $E(b_2)$, ... are all nonnegative. A Dutch book is impossible. Judgments that are mathematical probabilities are coherent:

Judgmental probabilities are coherent if and only if they have the mathematical structure of classical probabilities.

UPDATING PROBABILITIES

We have answered our initial question, but once one begins thinking in this way, it leads further into interesting territory. We have shown something about static coherence—about judgmental probabilities that are coherent at a fixed time. Now we will be able to show something about dynamic coherence—coherent belief change across time. First we have to understand something that de Finetti shows us about conditional bets.

CONDITIONAL BETS

De Finetti had an additional important idea about coherence, one that relates conditional probabilities to conditional bets.[6] A conditional bet is one that is called off provided that the condition isn't met. So a bet on q conditional on p has the form

$$\$a \text{ if } p \text{ and} \quad (\text{bet won if } p \text{ and } q),$$

$$\$-b \text{ if } p \text{ and not } q \quad (\text{bet lost if } p \text{ and not } q),$$

$$0 \text{ otherwise} \quad (\text{bet called off if not } p).$$

The imputed *conditional probability*, $P(q\,|\,p)$, of a conditional bet judged as *fair* is $b/(a+b)$.

Example: Fred's probability that Sarah will be nominated and she will win is $\frac{1}{3}$. He considers a bet that pays $\frac{2}{3}$ if she is nominated and wins and loses $\frac{1}{3}$ otherwise as fair. Fred's probability for Sarah being nominated is $\frac{1}{2}$. He considers a bet that loses $\frac{1}{3}$ if she is nominated and wins $\frac{1}{3}$ otherwise as fair. Notice that these two bets taken together constitute a conditional bet on Sarah winning conditional on her being nominated, with an imputed conditional probability of $\frac{2}{3}$. (If she is not nominated, he wins $\frac{1}{3}$ and loses $\frac{1}{3}$ for a net payoff of 0. This is just like calling off a bet if she is not nominated.)

Now suppose his conditional betting rate disagrees with this conditional probability. For instance, suppose he considers even money bets conditional on her nomination as fair, and acts accordingly. Then he is incoherent, and open to a Dutch book, described as follows.

In order to make everything transparent, we will make the Dutch book in two stages.

Stage 1: First we make a bet, B_1, where we pay Fred $\frac{2}{3}$ if Sarah is nominated and wins and he pays us $\frac{1}{3}$ otherwise. Next we make Fred a bet B_2, where he pays us $\frac{1}{3}$ if she is nominated, and we pay him $\frac{1}{3}$ otherwise. Finally, we make a bet on winning conditional on being nominated, B_3, where Fred pays us $\frac{1}{2}$ if she is nominated and wins, we pay him $\frac{1}{2}$ if she is nominated and loses, and the bet is called off if she isn't nominated. The following table shows what he gets in each possible circumstance. This is only a conditional Dutch book; Fred has a sure loss only if Sarah is nominated.

Nominated	Elected	B_1	B_2	$B_1 + B_2$	B_3	Total
T	T	$\frac{2}{3}$	$-\frac{1}{3}$	$\frac{1}{3}$	$-\frac{1}{2}$	$-\frac{1}{6}$
T	F	$-\frac{1}{3}$	$-\frac{1}{3}$	$-\frac{2}{3}$	$\frac{1}{2}$	$-\frac{1}{6}$
F	T	$-\frac{1}{3}$	$\frac{1}{3}$	0	0	0
F	F	$-\frac{1}{3}$	$\frac{1}{3}$	0	0	0

Stage 2: To turn it into a full Dutch book, we hedge. We do this by making an additional even money bet, B4, according to which we pay Fred $\frac{1}{12}$ if she is nominated and he pays us $\frac{1}{12}$ if she isn't. He considers this as fair. His net loss is now $\frac{1}{12}$ no matter what. Fred is subject to a Dutch book. If you want to see the general case worked out, it is in an appendix to this chapter.

P	Q	B_1	B_2	$B_1 + B_2$	B_3	B_4	Total
T	T	$\frac{2}{3}$	$-\frac{1}{3}$	$\frac{1}{3}$	$-\frac{1}{2}$	$\frac{1}{12}$	$-\frac{1}{12}$
T	F	$-\frac{1}{3}$	$-\frac{1}{3}$	$-\frac{2}{3}$	$\frac{1}{2}$	$\frac{1}{12}$	$-\frac{1}{12}$
F	T	$-\frac{1}{3}$	$\frac{1}{3}$	0	0	$-\frac{1}{12}$	$-\frac{1}{12}$
F	F	$-\frac{1}{3}$	$\frac{1}{3}$	0	0	$-\frac{1}{12}$	$-\frac{1}{12}$

COHERENT UPDATING

So far we have coherence of conditional and unconditional bets at a given time. What about change in probabilities when we get new evidence? Is there a sense of coherent belief change that applies? Suppose that the evidence is some proposition, e, which you learn with certainty. Then the standard rule for changing one's judgmental probabilities is to take as one's new probabilities the old probabilities conditional on e. This is known as *conditioning on the evidence*. Is there a coherence argument for this rule? We want to emphasize that we have shifted from asking about coherence of degrees of belief to asking about *coherence of rules for changing degrees of beliefs*.

Such an argument is implicit in de Finetti's discussion, and it was made explicit by the philosopher David Lewis.[7] Conditioning on the evidence, Bayesian updating, is the unique coherent rule for updating probabilities in such a situation. Any other rule leaves one open to a Dutch book against the rule—a Dutch book across time, a *dia-chronic Dutch book*. Here are a model and a precise version of the argument.

The Model

The epistemologist (scientist, statistician) acts as bookie. She has a principled way of updating on the evidence, but we don't presuppose what it is. *Today* she posts her probabilities and does business. *Tomorrow* she makes an observation (with a finite number of possible outcomes, each with positive prior probability.) She updates her probabilities according to her *updating rule*, a *function* that maps possible observational outcomes to revised probabilities. Her updating rule is public knowledge. The day after tomorrow—after her observation—she posts revised fair prices and does business.

A bettor's strategy consists of (1) a finite number of transactions today that the epistemologist considers fair according to her probabilities, and (2) a function taking possible observations to sets of finite transactions the day after tomorrow at the prices the epistemologist *then* considers fair according to her updating rule.

COHERENT BELIEF CHANGE IMPLIES CONDITIONING ON THE EVIDENCE

Let $P(A|e)$ be $P(A$ and $e)/P(e)$ and $P_e(A)$ be the probability that the bookie's nonstandard updating rule—the rule at variance with conditioning on the evidence—gives A if e is observed. Suppose $P(A|e) > P_e(A)$[8] and let the discrepancy, δ, be $P(A|e) - P_e(A)$. Here is the bettor's strategy that makes a Dutch book.

Today: Offer to sell the bookie at her fair price:

1. [\$1 if A and e, 0 otherwise].
2. [\$$P(A|e)$ if not e, 0 otherwise].
3. [\$$\delta$ if e, 0 otherwise].

Tomorrow: If e was observed, offer to buy [\$1 if A, 0 otherwise] from the bookie at its current fair price: $P_e(A) = P(A|e) - \delta$.

Then, in every possible situation, the bookie loses \$$\delta P(e)$.[9]

We could, of course, take a little of our sure winnings and divide them up to sweeten each transaction. *Then the bookie finds every transaction leading to the Dutch book to her advantage at the time she makes it.*

CONDITIONING ON THE EVIDENCE IMPLIES
COHERENT BELIEF CHANGE

If the epistemologist updates by conditioning on the evidence, then every payoff that the bettor's strategy can achieve can be gotten by only betting today (using conditional bets). So if her judgmental probabilities today are coherent and she updates on the evidence, a Dutch book cannot be made against her.

COHERENT BELIEF CHANGE AS CONSISTENCY

But can we say somehow that this vulnerability to a diachronic Dutch book is a result of giving inconsistent evaluations to two ways of doing the same thing? The bets in Lewis's Dutch book should seem familiar. The first two are just the ways of making a conditional bet out of two unconditional bets. The third is a little side bet. This looks just like de Finetti's Dutch book argument for conditional probability.

Let's say that we already have de Finetti's argument in the bank and know about consistent evaluation of conditional bets. Then the diachronic Dutch book just comes to this—*we have two different ways of making a conditional bet within our epistemic model.* One is to make a conditional bet today in the way that de Finetti showed us. Another is to wait until tomorrow and bet at tomorrow's rates just in case the condition—that evidence *e* came in—has been satisfied. *If the two agree, we have changed our judgmental probabilities by conditioning on the evidence.* Coherent belief change coincides with conditioning on the evidence.

COHERENT UPDATING EXTENDED

The Dutch book for conditioning on the evidence that we just presented took place within a specific, rather stylized, epistemic model. One could loosen the model or vary it in any number of ways. There is, in fact, a literature in which both statisticians and philosophers investigate coherence in a variety of more-or-less structured epistemic models. A good place to start is Freedman and Purves, "Bayes Method for Bookies"[10]—in which it is not assumed that the statistician has any prior probabilities at all, and in which it is shown that if she is coherent she will act *as if* she does have prior probabilities and conditions on the evidence.

There are extensions to cases where the evidence is uncertain. In Richard Jeffrey's *probability kinematics*,[11] there is no evidential proposition that we learn with certainty. Rather, the evidential experience causes shifts in the probabilities of evidential propositions while leaving probabilities of other propositions conditional on members of the evidential partition unchanged. This leads to a rich general conception of updating with connections to minimal change of probabilities.[12] There is an appropriate sense of coherence and a Dutch book here as well.[13] If you find this interesting, see appendix 2.

Mathematically, Dutch book theorems are part of the theory of arbitrage. The market plays the role of bookie, buying and selling at the going price. An incoherent market affords opportunities to buy and sell the same thing at different prices, affording opportunities for the arbitrager to exploit inconsistencies in the market and pocket a profit. If the market is one for predictions, and there are many such markets, then an arbitrage is a Dutch book. The theory of arbitrage can be applied to give a quite general theory of coherence both *at* a time and *across* time.[14]

Since some people feel uncomfortable with a gambling justification of rational belief, de Finetti provided an alternative justification, based on *calibration*.[15, 16] We are interested in having true beliefs. Give truth a value 1 and falsehood a value 0. Before finding out the truth, you have some degree of belief. Suppose that after finding out the truth, you are assessed a squared error penalty. For example if you take $P(A)$ to be 0.9 and A turns out true, your penalty is 0.01, but if A turns out to be false, your penalty is 0.81.

De Finetti shows that if you are incoherent, there is a coherent degree of belief (a probability) that will incur a smaller penalty no matter how the truth falls out. Consider a simple example. Suppose that A and B are mutually exclusive. Then there are only three possibilities for truth: both false; A true, B false; A false, B true. Probabilities are weighted averages[17] of truth values. Letting $x = P(A)$, $y = P(B)$, and $z = P(A \text{ or } B)$, these form the two-dimensional object $z = x + y$ shown in figure 2.1.

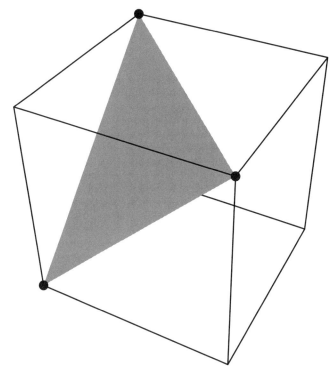

Figure 2.1. Possible probabilities for three alternatives

The coherence arguments show that these coincide with the coherent degrees of belief. Suppose that some degrees of belief are here incoherent. Then they will correspond to a point *off* the surface. Euclidian geometry assures that there is a point *on* the surface that is closer to the truth, no matter what the truth. These scoring rules have a variety of implications. They have been used to keep weather forecasters calibrated and to help keep an expert's expressed opinion in line with the expert's honest opinion.[18]

Although de Finetti's two justifications of judgmental probability seem to have a quite different flavor, they both turn on the very same mathematical property[19]—weighted averages of probabilities are probabilities. If degrees of belief are not weighted averages of truth values—probabilities 1 or 0—bad things happen. One bad thing is being subject to a Dutch book; another is surely being further from the truth than one could have been if coherent.

PART II: UTILITY AND JUDGMENTAL PROBABILITIES

Money isn't everything. In most human affairs payoffs are not in terms of money. They are in terms of whatever we value. With real goods, the values of payoffs from two bets may not add. There are complementarities. Two goods held together may be such that each enhances the value of the other or conversely. Even money may not satisfy the assumptions of a Dutch book because of risk aversion. If one bet hedges another, it may be complementary because it reduces risk. In modern terms, utility is not linear with respect to money. The answer to these worries is to reformulate the theory not in terms of money, but in terms of utility.

But how is utility to be measured? And can probability and utility both be measured in a noncircular way? The answer is affirmative, and it was given by a man of genius, Frank Plumpton Ramsey, in a remarkable essay, "Truth and Probability." Our story, however, begins much earlier and takes a few turns before we get to Ramsey.

UTILITY

Wise men have always known that money is not the true measure of value, but the point was focused on the theory of gambling by Nicholas Bernoulli's "St. Petersburg game" and its subsequent analysis. Writing to Montmort[20] on September 9, 1713, Nicholas Bernoulli poses some problems, including the following two ("Correspondence of Nicolas Bernoulli Concerning the St. Petersburg Game," tr. Richard J. Pulskamp. Available online at http://cerebro.xu.edu/math/Sources /NBernoulli/correspondence_petersburg_game.pdf.):

> Fourth Problem. A promises to give a coin to B, if with an ordinary die he achieves 6 points on the first throw, two coins if he achieves 6 on the second throw, 3 coins if he achieves this point on the third throw, 4 coins if he achieves it on the fourth and thus it follows; one asks what is the expectation of B?
>
> Fifth Problem. One asks the same thing if A promises to B to give him some coins in this progression 1, 2, 4, 8, 16 etc. or 1, 3, 9, 27 etc. or 1, 4, 9, 16, 25 etc. or 1, 8, 27, 64 instead of 1, 2, 3, 4,

5 etc. as beforehand. Although for the most part these problems are not difficult, you will find however something most curious.

Montmort replies that there is really no difficulty in solving these— simply sum the infinite series: "the late Mr. your uncle has given the method to find the sum of these series" (referring to Jacob Bernoulli).

Nicholas's rejoinder is that Montmort should have actually tried it. Although the series for the fourth problem sums to six, those in the fifth problem sum to infinity. This is the "something most curious." What does this mean? How could this gamble have a value greater than any finite sum? Both are puzzled, and Nicholas poses the problem to others. On May 17, 1728, Gabriel Cramer, a Swiss mathematician writes to Nicholas from London ("Correspondence of Nicolas Bernoulli Concerning the St. Petersburg Game," tr. Richard J. Pulskamp. Available online at http://cerebro.xu.edu/math/Sources/NBernoulli/correspon dence_petersburg_game.pdf.):

> I know not if I deceive myself, but I believe to hold the solution of the singular case that you have proposed to Mr. de Montmort . . . . In order to render the case more simple I will suppose that A throw in the air a piece of money, B undertakes to give him a coin, if the side of the Heads falls on the first toss, 2, if it is only the second, 4, if it is the 3rd toss, 8, if it is the 4th toss, etc. The paradox consists in this, that the calculation gives for the equivalent that A must give to B an infinite sum, which would seem absurd, since no person of good sense, would wish to give 20 coins. One asks the reason for the difference between the mathematical calculation and the vulgar estimate. I believe that it comes from this that the mathematicians estimate money in proportion to its quantity, and men of good sense in proportion to the usage that they may make of it.

Cramer goes on to make two crucial points. The first is that if the bank is not infinite, the game can only go up to some fixed limit, and even if this is large, the expected utility of the gamble will be moderate. The second, which is of interest here, is that true value is not proportional to money:

One will be able to find again a smaller (value) by making some other assumption of the *Moral Value* of the riches. Because what I come to make is not exactly just, since it will be true that 100 millions are more pleasure than 10 millions, although not making 10 times more.

Here we have the idea of utility—*moral value*—clearly distinguished from monetary gain. Ten years later, Daniel Bernoulli—Nicholas's cousin—published the problem and essentially the same solution in *Commentaries of the Imperial Academy of Science of Saint Petersburg*, thus the Saint Petersburg problem. Daniel had learned of Cramer's work from Nicholas and gives Cramer full credit (Daniel Bernoulli, "Exposition of a New Theory on the Measurement of Risk," tr. Louise Sommer, *Econometrica* 22 (1954: 23–36): 33):

> Then this distinguished scholar informed me that the celebrated mathematician, Cramer, had developed a theory on the same subject several years before I produced my paper. Indeed I have found his theory so similar to mine that it seems miraculous that we independently reached such close agreement on this sort of subject.

Cramer and Daniel Bernoulli suggested specific utility functions: Cramer considered utility as the square root of money. Bernoulli assumed that the utility generated by an increment in money should be inversely proportional to the fortune you already have, and derived utility of your fortune being equal to the logarithm of its size.[21] Bernoulli goes on to develop an account of risk aversion and to discuss the rationality of buying insurance.[22, 23]

MEASURING UTILITY

It makes no sense to talk about utility as a function of money unless utility is a quantity. How do you measure utility? The English Utilitarians of the nineteenth century regarded this as a problem to be solved by psychology—or perhaps by a combination of psychology and philosophy. They agreed with Cramer and Bernoulli that utility of an additional unit of money declines as one gets richer and take this to be a basis for social reform.

But in the early twentieth century, a positivist school of economics pressed the question. If utility could not be measured, then the only utility statements with empirical content were comparative and subjective, such as: For Karl, utility of *A* is greater than utility of *B*, and this just meant Karl prefers *A* to *B*, or better, if given the choice, Karl will choose *A* over *B*. Utility scales had only ordinal significance—that is to say that any two utility scales with the same ordering have the same empirical content.

This all changed when John von Neumann and Oskar Morgenstern published *Theory of Games and Economic Behavior* in 1944.[24] There a cardinal theory of utility is constructed on ordinal foundations. This is done by bringing in classical chance. We will show how.

Suppose that according to your preferences, some outcome—call it GOOD—is the best and another, call it BAD, is the worst. We suppose that you have preferences over gambles (or lotteries) of the form: GOOD with *chance p*, BAD otherwise. We can choose the utilities of GOOD and BAD arbitrarily, so long as the former is larger than the latter. This is just the choice of 0 and unit in the scale on which utility is measured (like Celsius and Fahrenheit scales of temperature). To make things simple, we choose utility of GOOD to be 1 and utility of BAD to be 0.

Now, for the utility of any other outcome, *O*, we find a gamble GOOD with *chance p*, BAD otherwise, so that you are indifferent between it and *O* for sure.* Then you take the utility of *O* to be equal to *p*. That is to say that the utility of *O* is equal to the expected utility of the gamble that you judge equally good. Starting with chance in hand, the expected utility principle is used to measure utility. Notice that the ordinalist has nothing to complain about here. Only ordinal judgments have been used here, although the scope of the judgments has been extended to objective gambles. It is the chance in the gambles that has put some quantitative backbone in the ordinal story, yielding cardinal utilities.

To be careful, we note that use of the term *gamble* may carry extraneous connotations. Here, by a gamble we just mean a probability distribution over outcomes. So what we might describe for imagery

---

*What are we assuming when we assume you can do this? Bear with us; it will be explained.

as a gamble over gambles is just another probability distribution over outcomes. We assume that an individual has consistent preferences over all such gambles over our finite set of consequences. Then it is a theorem that there exists a *utility* such that expected utility of gambles agrees with her preferences.

Let us spell out the requirements for preferences in a little more detail. First the preferences must totally order the gambles. This does not mean that of two gambles, you have to prefer one to another; there can be ties. But one cannot be entirely flummoxed by the comparison, not being able to say whether A is better than B, or worse, or just as good. This is an idealization for an ideal set of preferences. Two further ideal properties that are assumed are continuity and independence.

Continuity: If p *preferred* p' *preferred* p'', then there is a probability, a, such that

$$ap + (1-a)p'' \text{ indifferent } p'.$$

Independence: p *preferred* p' if and only if

$$ap + (1-a)p'' \text{ preferred } ap' + (1-a)p'', \text{ for every } a \text{ and every } p''.$$

In the second line of the statement of *independence*, both gambles give you p'' with probability $(1-a)$. They only differ in what they give you with probability a. So your preference should be controlled by your preference for what they give with probability a, which was stated in the first line. That is the content of independence.

Without doing the full argument, we can give a taste of how these properties come into play. First, we make a utility scale for gambles just between GOOD and BAD, as explained before. Numerical utility is just the probability of GOOD. Now we need to show that for these gambles between GOOD and BAD, utility represents your preference ordering among gambles. We know you prefer GOOD to BAD, but we must show, for instance, that you prefer GOOD to a nontrivial gamble between GOOD and BAD to BAD.

Consider a nontrivial gamble, p GOOD $+ (1-p)$ BAD.

GOOD p GOOD $+ (1-p)$ BAD   BAD

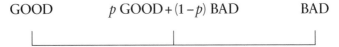

Since GOOD is equivalent to GOOD for sure, BAD is equivalent to BAD for sure, and GOOD is preferred to BAD, we can apply independence,

$$p \text{ GOOD} + (1-p) \text{ GOOD},$$

$$p \text{ GOOD} + (1-p) \text{ BAD},$$

$$p \text{ BAD} + (1-p) \text{ BAD},$$

to conclude that preferences line up as they should. Slightly more complicated application of independence shows this for two nontrivial gambles; the higher gamble on the utility scale is preferred to the lower.

Now the utility of any gamble, q, other than our special ones between GOOD and BAD is measured by finding one of the special gambles, p, such that one is indifferent between it and q. We then assign the gamble q the same utility. Why can we always find such a reference gamble? This is assured by continuity. The Von Neumann–Morgenstern theorem for a finite number of consequences is completed by repeated application of independence, ordering, and continuity.

Philosophical worries about the use of money in the Dutch book arguments that were noted in the last chapter all vanish if we take our payoffs to be in terms of von Neumann–Morgenstern utilities. But aren't we now in a circle? We used utilities to measure judgments of chance, and now we are using judgments of chance to measure utilities.

It is not quite a circle. We could help ourselves to classical chance devices—the dice, the fair lottery, the wheel of fortune—and use them to measure an individual's utilities of everything. Then we could use these utilities to measure her judgmental probabilities for things other than outcomes of classical chance devices—whether it will rain tomorrow, the outcome of the election, whether everyone will be laid off at work. This was all worked out in a simple and elegant way by Anscombe and Aumann in 1964.[25]

We are, however, helping ourselves to the classical equally probable cases and stipulating that the agent in question takes them to be equally probable. A thoroughgoing judgmental account would get all probabilities *and* all utilities out of personal preferences. Now—you might be tempted to say—this really is impossible. But this is just what was accomplished by Frank Plumpton Ramsey in his essay "Truth and Probability."[26] How was it done?

RAMSEY

We already have some of the pieces of the puzzle. The one that is still missing is Ramsey's idea of an "ethically neutral" proposition. This is a proposition, p, whose truth or falsity, in and of itself, makes no difference to an agent's preferences. That is to say, for any collection of outcomes, B, the agent is indifferent between B with p true and B with p false. How about whether the number of milliseconds since I turned on my computer is odd or even? How about whether the coin flip comes up heads or tails? Typically, we care only about some things, and we have lots of ethically neutral propositions. Intuitively, the nice thing about gambles on ethically neutral propositions is that the expected utility of gambles on them depends only on their probability and the utility of the outcomes. Their own utility is not a complicating factor.

Now we can identify an ethically neutral proposition, h, with probability $\frac{1}{2}$ as follows. Consider two outcomes, A, B, such that you prefer the first to the second. Then the ethically neutral proposition, h, has probability $\frac{1}{2}$ for you if you are indifferent between [A if h; B otherwise] and [B if h; A otherwise]. Now we have identified a subjective surrogate for a fair coin toss. This is the key idea. We can use this over and over to construct our utility scale, which is what Ramsey does. This is a slight variation on the von Neumann–Morgenstern procedure, done much earlier than von Neumann and Morgenstern. Now that we have utilities of outcomes, we can use them to measure probabilities of propositions that are *not* ethically neutral, as with de Finetti. Let's see how this works in an example.

The Horse Race

Consider four propositions HH, HT, TH, TT, which are mutually exclusive and jointly exhaustive. Farmer Smith doesn't really care which of these is true. More specifically, for whatever way the things he does care about could come out, he is indifferent to their coming out that way with HH, or with HT, or with TH, or with TT. Then, in Ramsey's terminology, these four propositions are *ethically neutral*.

Suppose, in addition, that for any things he does care about, *A* preferred to *B* preferred to *C* preferred to *D*, he is indifferent between the gamble

$$A \text{ if HH,}$$

$$B \text{ if HT,}$$

$$C \text{ if TH,}$$

$$D \text{ if TT,}$$

and any other gamble that can be gotten by rearranging *A*, *B*, *C*, *D*, for instance,

$$D \text{ if HH,}$$

$$B \text{ if HT,}$$

$$C \text{ if TH,}$$

$$A \text{ if TT.}$$

Then for him, HH, HT, TH, TT all have the *same probability*, equal to $\frac{1}{4}$. (Perhaps this is because these events represent what he takes to be two independent flips of a fair coin, and he is making judgments like Cardano, Galileo, Pascal, and Fermat.)

There is to be a race in which the horses, Stewball and Molly, compete. Farmer Smith owns Stewball, and the propositions Stewball wins and Molly wins are not at all ethically neutral for him. He can wager on the race, with the prospect of winning a pig if the horse he wagers on wins.

His most preferred outcome is get pig and Stewball wins, and he assigns it utility equal to 1. His least preferred outcome is no pig and Stewball loses, and he assigns it a utility of 0. These are just arbitrary choices of how to choose his utility scale:

1 | Get Pig and Stewball Wins,

.
.
.

0 | No Pig and Stewball Loses.

Farmer Smith is indifferent between: get pig and Molly wins and a hypothetical gamble that would ensure that he would get the pig and Stewball would win if HH or HT or TH and that he would get no pig and Stewball would lose if TT. But that gamble has expected utility $\frac{3}{4}$, so we can begin filling in our utility scale:

$1 \cdot$ Get Pig and Stewball Wins,

$\frac{3}{4} \cdot$ Get Pig and Molly Wins,

.
.
.

$0 \cdot$ No Pig and Stewball Loses.

He is indifferent between no pig and Molly loses and the hypothetical gamble that would ensure that he would get the pig and Stewball would win if HH and that he would get no pig and Stewball would lose if HT or TH or HH. Now we have

1 | Get Pig and Stewball Wins,

$\frac{3}{4}$ | Get Pig and Molly Wins,

.

$\frac{1}{4}$ | No Pig and Molly Loses,

0 | No Pig and Stewball Loses.

He is indifferent between the gamble pig if Molly wins and no pig if she loses and the gamble get pig and Stewball wins if HH or HT, but no pig and Stewball loses if TH or TT. The first gamble is not conditioned on ethically neutral propositions, but it is equated to 1; that is, the gamble get pig and Stewball wins if HH or HT, but no pig and Stewball loses if TH or TT has expected utility $\frac{1}{2} \cdot 1 + \frac{1}{2} \cdot 0 = \frac{1}{2}$. So the first gamble, pig if Molly wins and no pig if she loses, must also satisfy

P(Molly wins) Utility(pig and Molly wins) $+ (1 - P$(Molly wins))

Utility(no pig and Molly loses) $= \frac{1}{2}$.

We know the two utilities already, $\frac{3}{4}$ and $\frac{1}{4}$, respectively, so we now have pinned down Farmer Smith's judgmental probability of a non-ethically neutral proposition. He takes Molly to be an even-money bet.

P(Molly wins) $= \frac{1}{2}$.

Ramsey started with a coherent preference ordering and showed how to extract probabilities and utilities such that preferences are in accord with expected utility. (This is only sketched in Ramsey's essay, but the key ideas are all there.) This is a representation theorem for probability and utility. Coherent preferences can be represented as coming from judgmental probability and personal utility by the rule of preferring more expected utility to less. If the preference structure is rich enough, then typically the probabilities are determined uniquely and the utilities up to the choice of 0 and 1.

This is even more remarkable in that it was done before Anscombe and Aumann, before von Neumann and Morgenstern, and even slightly before de Finetti. Very careful versions of such representations for probability were later worked out, most notably by L. J. Savage in *The Foundations of Statistics* in 1954.[27]

Of course, the assumptions going into these utility-probability representation theorems are highly idealized. None of the great contributors were under any illusions with regard to this. Ramsey, for instance, wrote (F. P. Ramsey, "Truth and Probability" (1926), in *The Foundations of Mathematics and Other Essays*, ed. R. B. Braithwaite (London: Routledge, 1954))

I have not worked out the mathematical logic of this in detail, because this would, I think, be rather like working out to seven places of decimals a result only valid to two. My logic cannot be regarded as giving more than the sort of way it might work.

But for many practical affairs, working out probability and utility to two decimal places would be very good indeed, and working them out to one place considerably better than not working them out at all.

We have seen the development of the view that probability theory is *logic*—the logic of coherent degrees of belief. In the final analysis, it is a kind of pragmatic logic. It is a logic of decision, a logic of coherent preferences over acts with uncertain outcomes. As such, judgmental probability for a coherent agent can be measured in terms of dispositions to choose. This leads us to ask the question of the next chapter—how coherent are typical human agents?

SUMMARY

Strength of judgment can be measured, and coherent judgmental strengths are probabilities.

Suppose we have a measure of value and buy and sell wagers according to their expected value. If the weights by which we form the expectation do not behave as mathematical probabilities, we can be made to play the fool (Dutch booked). If they behave as probabilities, we cannot be made to play the fool. Versions of the same argument can be can be brought to bear on updating probabilities in the light of new evidence.

If we do not already have a measure of value, we can make one and simultaneously make a model of strengths of judgment starting from coherent preferences such that

– Strengths of judgment are mathematical probabilities,
– Values are utilities, and
– The original order of preference agrees with expected utility.

APPENDIX 1. COHERENCE OF CONDITIONAL BETS

For arbitrary propositions *p*, *q*, consider bets on *p* and *q* and against *p* and the result of holding them together, as shown in the table.

p	q	Bet on p and q	Bet on not p	Combination
T	T	c	$-f$	$c-f$
T	F	$-d$	$-f$	$-(d+f)$
F	T	$-d$	e	$e-d$
F	F	$-d$	e	$e-d$

If the stakes on the bet against p are chosen such that $d=e$, the result is equivalent to a conditional bet.*

p	q	Bet on p and q	Bet on not p	Combination
T	T	c	$-f$	$c-f$
T	F	$-d$	$-f$	$-(d+f)$
F	T	$-d$	d	0
F	F	$-d$	d	0

Suppose that the individual bets and the conditional bet are all judged as fair. Then the imputed probabilities are

$$P(p \text{ and } q) = \frac{d}{c+d},$$

$$P(p) = \frac{d}{f+d},$$

$$P(q \text{ given } p) = \frac{d+f}{d+c}.$$

That is to say, $P(p \text{ and } q) = P(p)P(q \text{ given } p)$ or that $P(q \text{ given } p) = P(p \text{ and } q)/P(p)$. This is the multiplication law for probability (usually presented as a definition of conditional probability).

If conditional bets and unconditional bets cohere, then the addition of conditional bets to our markets adds nothing new. Any payoffs achieved by using conditional bets could have also been achieved by using the equivalent set of unconditional bets. So a Dutch book cannot be made. If, on the other hand, an individual's judgmental probabilities lead to inconsistent evaluations of the two ways of achieving a conditional bet, we can obviously make a Dutch book by

---

*This can always be done, provided that there is a positive probability imputed to the condition.

buying cheap and selling dear. *Coherence is again a question of consistent evaluations of a betting arrangement that can be implemented in alternative ways.*

APPENDIX 2. PROBABILITY KINEMATICS

Suppose you get up at night and observe, by the dim light of the moon coming in through a window, a jellybean sitting on a table. Jellybeans that might possibly be there are red, pink, brown, or black. The light it not good enough to tell the color exactly, but it is good enough to shift your probabilities. This is a case of uncertain observation; there is no proposition available to sum up the content of your observation. You might try saying that the proposition is just *this* experience, but that is not a proposition in any reasonable probability space that we can profitably think about. This is an example of uncertain evidence.

This may seem to be an example not worth serious thought. But we are often confronted with uncertain evidence. Important decisions may be made based on observation by candlelight or by the light of the moon. What did that smile mean? Or, on a more sober note, think of a radiologist reading a scan of a lung or a pathologist interpreting the results of a biopsy.[28] If you think a little harder, uncertain evidence is all around us.

We often deal with it by pretending that we have certain evidence. An interpretation is made. A proposition is treated as certain, when it isn't. Sometimes that is good enough. But, you will agree, the question of uncertain evidence is worthy of serious thought.

Back to our toy example: jellybeans have different flavors. A red one may be cherry or cinnamon. A brown one may be chocolate or coffee. There are all sorts of jellybeans in your house. Most of the brown ones are chocolate, but there are a few coffee ones. We suppose you have a probability for a brown one being chocolate, and likewise for the other colors and flavors. Then if you just looked at the jellybean and didn't taste it, it would seem reasonable for the probabilities of taste conditional on color to remain unchanged while the probability distribution among the colors shifts. If so, your belief change has been by what Richard Jeffrey called *probability kinematics* on the colors. (In the pathology example of Jeffrey and Hendrickson, note 28, diagnosis takes the place of colors and prognosis the place of flavors.)

In what sense can considerations of coherence be brought to bear on probability kinematics? We elaborate our toy model. Suppose that you have a probability over color-flavor pairs before the observation (P_1) and one after the observation (P_2). Suppose that after that, someone—also up—walks in the room and turns on the lights. Then you have a third probability (P_3). And we assume that now you are certain of the color.

Now we need a way of saying that your observational experience by moonlight is *just about colors*—that you didn't taste the jellybean. We can do this by postulating that the certain evidence that you get by observing the color when the light is turned on swamps the uncertain evidence that you previously got. That is to say, your final probabilities are the same as those you would have gotten if the light had just been turned on and you observed color, without any mediating uncertain observation of color. This is a way of saying that colors form a *sufficient partition* for your uncertain evidence.

To bring coherence to bear, assume that you have a coherent rule for updating on *certain* evidence. By the argument given in the chapter, it must be the rule of conditioning on the evidence. Then, by coherence, P_3 comes from P_2 by conditioning on the color observed. By coherence and sufficiency, P_3 comes from P_1 by conditioning on the color observed. It follows that P_2 comes from P_1 by probability kinematics on the partition of colors (since conditioning on a color leaves probabilities of flavors conditional on that color unchanged).[29]

Amos Tversky

CHAPTER 3

———

PSYCHOLOGY

Our third great idea is that *the psychology of chance and the logic of chance are quite different subjects.* The ideal normative assumptions of chapter 2 are often violated in practice. Normative and descriptive aspects of judgmental probability and decision theory come apart, leaving a substantial gulf.

Of course, there has been an appreciation for a long time that the subjects are not *exactly* the same. But there has been a widespread feeling that they are close enough for many applications. Frank Ramsey described his theory as exact logic but also approximate psychology. The dominant stance in economics, at least until recently, has been just the same. Expected utility theory may not be exact psychology, but it is good enough for an empirical science to use as its core theory of decision-making.

The paradigm that identified (or almost identified) psychology and logic was so strong that those who advanced the first robust major deviations of psychology from expected utility theory, Maurice Allais and Daniel Ellsberg, reacted by trying to transform the logic to match the psychology.[1] There is no question but that the attempt to do so led to interesting theoretical work. But the case that the gap is unbridgeable was made with great force by Daniel Kahneman and Amos Tversky in a whole series of studies. As evidence piled up, the doors were opened to the application of empirical psychology to empirical economics. This has come to full flower in the current flourishing field of behavioral economics.[2]

Behavioral economics can have implications for public policy. If individuals do not behave as rational actors, policies that assume that they do may have unfortunate consequences. One important area where deviations from rational decision models can occur is in the area of health care policy. Individuals tend to neglect low-probability, high-risk events in the future and to overweight costs in the present. In a policy analysis, Liebman and Zeckhauser argue that, as a result, there is a tendency to undersubscribe to health care insurance, although conventional economic theory predicts a tendency to oversubscribe to subsidized insurance.[3] They conclude that "policy analyses of the wide range of subsidies that permeate the health care system change substantially when viewed from the behavioral perspective" (Jeffrey Liebman and Richard Zeckhauser, "Simple Humans, Complex Insurance, Subtle Subsidies," Working Paper 14330; http://www.nber.org/papers/w14330, National Bureau of Economic Research, 1050 Massachusetts Avenue, Cambridge, MA 02138 (September 2008), 1).

If the psychology of the consumer of health care, or those impacted by social services more generally—public education, or the criminal justice system—departs from rational choice, then social policy might well pay attention to psychology. Here we give an introduction to the contrast between the psychology of decision and the logic of decision.

In the last chapter we saw how Ramsey, in 1926, sketched the idea of a probability-utility representation of preferences. If a person's preferences over a sufficiently rich set of alternatives satisfied certain plausible principles, they could be represented as arising from personal probabilities and personal utilities *via* expected utility—options with higher expected utility preferred to those with less.

Daniel Kahneman

Representation theorems of this type were later proved, notably by Anscombe and Aumann, who assumed the availability of objective probability, and Savage, who like Ramsey, did not. What are these "certain plausible principles"? We met two key players in our discussion of von Neumann–Morgenstern utility in the last chapter. These are *ordering* and *independence*. Ordering says that you really know how to choose. Either you prefer p to q or q to p or you are indifferent between them. And your preferences are transitive. Independence says that your preferences between two lotteries should depend on only those outcomes in which the prizes differ. Ordering and independence appear in some form in the assumptions of both Savage and Anscombe and Aumann.

Savage sees independence (or sure-thing reasoning) as a fundamental principle of rational decision. But it was immediately challenged by the French economist Maurice Allais.[4] Allais set the following problems to Savage over lunch in 1952 at a conference on risk that he (Allais) had organized in Paris.

1. Choose between

    A. $1 Billion for sure.   B. $1 Billion with chance 89%,
                                        Nothing with chance 1%,
                                        $5 Billion with chance 10%.

2. Choose between

    A. Nothing with chance 89%,   B. Nothing with chance 90%,
        $1 Billion with chance 11%. $5 Billion with chance 10%.

(Note: The problem was originally framed in terms of millions rather than billions, but money isn't what it used to be.)

Take a minute to ask yourself how you would choose.

Savage chose A in the first problem and B in the second. Allais subsequently put the problem to other experts and many (but not all[5]) chose as Savage did. The experiment has been repeated many times, and the results are quite robust.

If you chose as Savage (and the authors of this book) did, take another minute and examine your reasons for your choice.

Now we will show how these choices violate independence. Think of these as choices between fair lotteries. There are tickets numbered from 1 to 100 and each has an equal chance of being drawn. Problem 1 can be redescribed as a choice between

A: $1 Billion on tickets 1–89   B: $1 Billion on tickets 1–89

+ $1 Billion on tickets 90–100   + $0 on ticket 90
                                  + $5 Billion on tickets 91–100.

Notice that the part of the wager above the dashed line is the same for A and B. So, by independence, any difference between them should be due to the part below the dashed line.

Let's delete the parts above the line and substitute payoff of 0 in each case:

A: $0 on tickets 1–89   B: $0 on tickets 1–89

+ $1 Billion on tickets 90–100   + $0 on ticket 90
                                  + $5 Billion on tickets 91–100.

This is just problem 2. Again, the parts above the line are the same for A and B, so by independence, choices should depend only on the part below. The parts below the dashed lines in the first and second problems are the same. If you choose A over B in the first problem, you should, by independence, choose A over B in the second problem.

Savage discusses the matter in *The Foundations of Statistics*. He says that when informed about the fact that his snap judgments violated the sure-thing reasoning, he went home, thought about it, and revised his judgments.

Suppose, like one of us, you chose A over B in the first problem on the basis that $1 billion is so much more than you would ever need and that the extra utility of money over that is essentially zero. Then you might like to reconsider your choices in the second problem on that basis and choose A over B there too.

Suppose, like many, you chose A over B in the first problem, thinking about how bad you would feel if you chose B and that unlikely unlucky ticket 90 came up. You would feel terrible; you would feel stupid; you would experience severe pangs of regret. If this is true, then the real payoff for ticket 90 in B is not zero but some negative number. *If you care about these feelings, they have to be counted as consequences.* Putting them in removes the violation of independence.

The Allais example, where many people (though not all) appear to violate plausible principles of rational preference, was soon joined by another set of somewhat different examples proposed by Daniel Ellsberg.[6]

Ellsberg—the same Daniel Ellsberg who leaked the Pentagon papers[7] and became an activist against the Vietnam War—proposed a set of problems designed to show the difference between choices involving *risk*, where objective probabilities are known, and those involving *uncertainty*, where they are not. In making the distinction and insisting on its importance, he follows the economists John Maynard Keynes[8] and Frank Knight,[9] who follow John Stuart Mill.[10] The distinction is drawn clearly in Ellsberg's first problem:

There are two urns. The first has 100 balls, some red and some black—you don't know the proportion. You know that the second has 50 red and 50 black balls.

Would you prefer $100 if a red ball is drawn from the first urn or $100 if a black ball is drawn? (Typically subjects are indifferent between these alternatives.)

This is a choice under uncertainty (or ambiguity).

Would you would prefer $100 if a red ball is drawn from urn 2 or $100 if a black ball is drawn? (The typical response is, again, indifference.) This is a choice under risk.

On the theory of subjective probability, someone making these choices assigns red and black subjective probabilities equal to $\frac{1}{2}$ in the case of urn 1 as well as that in the case of urn 2. But now a third choice is proposed: $100 on a red ball from the first urn or $100 on a red ball from the second urn?

Some subjects are indifferent. But many have a strong preference for the second urn—the one with known composition. The same preference for the second urn is manifest if the black balls are the ones that pay off. This is a choice *between* uncertainty and risk.

If you would make these choices, stop for a minute and examine your reasons.

How strong is your preference? Would you pay something to choose from the urn with known composition rather than the urn with unknown composition?

A strong preference for the second urn seems incompatible with the principle that preferences go by expected utility. Subject's within-urn preferences indicate that they regard red and black as equiprobable in each case. And probabilities sum to 1. But then a draw from the first urn should have the same expected payoff as one from the second.

Thus choices often made in Ellsberg's problem must violate at least one of Savage's axioms, but which ones are violated? We can narrow it down to either *ordering* or *independence*. If you are interested in developing the Ellsberg example to see which is violated, the appendix to this chapter shows how.

If you made the choices that violate the Ellsberg examples, as you reflect upon them, do you come up with a reason? Some people have a nagging feeling that they may somehow be cheated in the case of uncertainty. They are not sure how they would be cheated, but they think that they would be cheated. If this is how you feel, then the problem is not really a test of the Savage axioms.

Some people just get an unpleasant feeling making decisions under uncertainty that they don't get making decisions under risk. Who is to say that they shouldn't? (Others might get a thrill making decisions under uncertainty that is bigger than the thrill making decisions under risk. Who is to say that they shouldn't?) But if such psychological payoffs enter into Ellsberg cases, then they too should be counted as part of the consequences. And when they are counted, we no longer have a counterexample to the Savage axioms.

HEURISTICS AND BIASES

In 1974, Amos Tversky and Daniel Kahneman published "Judgment under Uncertainly: Heuristics and Biases."[11] Here they gave a panoply of reproducible deviations from subjective utility theory. They, and others, multiply the list in subsequent work. Their position is that snap judgments in this domain, as in other psychological domains, are guided by heuristics. These heuristics are often right and are thus very useful, but in certain circumstances they generate "cognitive illusions" that lead people astray. With time, effort, and knowledge of bias, errors can be corrected:

> A better understanding of these heuristics and of the biases to which they lead could improve judgments and decisions in situations of uncertainty. (Tversky and Kanhneman, p. 1131)

The article describes biases arising from three heuristics: *representativeness, availability*, and *adjustment and anchoring*. One version of the first is essentially judgment by stereotype. It is rapid, generally good as one's stereotypes, and liable to lead one to ignore evidence at variance with it. For instance an individual is described to conform to a common image of an engineer. If he is described as having been selected at random from either many engineers and a few doctors or from many doctors and a few engineers, it makes no difference to judgments about which he is. Stereotypes crowd out base-rate information. More careful reflection would use that information.

The second heuristic judges probability by how many readily available instances one can think of. This brings in all the vicissitudes of memory. Vividness, for instance, can bias judgments of probability. A newscast of a brush fire accompanied by dramatic video may increase my probability of brush fire, even if the video is from Australia. If something has been encountered recently (or something related has been encountered recently) it may get a higher estimated probability. This heuristic is systematically exploited by terrorists.

The third heuristic is anchoring. We start from an initial figure (the anchor) and then adjust to get a final estimate. Someone who can suggest the initial figure can influence the final estimate. The anchoring and adjustment effect is exploited by every used-car salesperson and every real estate agent. Exaggerated anchors are sometimes taken to great extremes in oriental bazaars, when bargaining is practiced as a fine art. One of the authors, while backpacking around the world, was bargaining in a bazaar and had beaten the price down to half the original asking price. He was about to settle, when a local friend said, "Wait, wait. That is too much," and negotiated a price that was *one-tenth* of the tentative half-price deal.

Psychologists know that bias due to lingering effect of an anchor can affect all kinds of quantities. Tversky and Kahneman show how insufficient adjustment from initial probabilities of two events can lead to overestimation of the probabilities of conjunctions and underestimation of the probabilities of disjunctions.

There are other examples in the paper and many more in Kahneman's book *Thinking Fast and Slow*.[12] The force of accumulating evidence made it less and less plausible to hold that subjective probability is, in general, approximate psychology.

FRAMING

In 1984, Kahneman and Tversky published "Choices, Values and Frames"[13] in *American Psychologist*. Here we concentrate on framing, which has direct application to the project of Allais and others to weaken the Savage axioms to get a decision theory that is descriptively adequate.

Consider their demonstration of *framing effects* in the following choices:

FRAMING I

The United States is preparing for an outbreak of a killer disease. With no action 600 people are expected to die. Choose between public health programs.

   1. Choose between the following:
     A. 200 people will be saved.
     B. Chance $\frac{1}{3}$ that 600 will be saved; chance $\frac{2}{3}$ that none will be saved.
   2. Choose between the following:
     A. 400 will die.
     B. No one will die with chance $\frac{1}{3}$, and 600 will die with chance $\frac{2}{3}$.

Kahneman and Tversky gave these questions, mixed with others, in a survey. They found that almost $\frac{3}{4}$ chose A over B in the first problem and almost $\frac{3}{4}$ chose B over A in the second. *But A and B mean the same thing* in both problems; only the wording is different.

These *framing effects*, where the choices made depend on the wording but not the substance of the decision, were found to be very common. Tversky and coworkers found them in life or death medical decisions made by doctors and patients, who would presumably be trying to be as careful as possible.[14] This suggests that medical decision making in general might be improved by practical applications of Bayes' theorem. This is a program that has been pursued by sophisticated physicians.[15] Bayes will be revisited in some detail in chapter 6.

Kahneman and Tversky asked subjects to choose a preferred treatment for lung cancer, with the problem alternately framed in terms of survival and of mortality.

FRAMING II

Survival Frame
   Surgery: Of 100 people having surgery, 90 live through the postoperative period, 68 are alive at the end of the first year, and 34 are alive after 5 years.
   Radiation therapy: Of 100 people having radiation therapy all live through the treatment, 77 are alive at the end of 1 year and 22 are alive after 5 years.

Mortality Frame

Surgery: Of 100 people having surgery, 10 die during surgery or the postoperative period, 32 die by the end of the first year, and 66 die by the end of 5 years.

Radiation therapy: Of 100 people having radiation therapy none die during treatment, 23 die by the end of 1 year, and 78 die by the end of 5 years.

In the survival frame, 18% favored radiation, while in the mortality frame, 44% favored radiation.[16]

Again, *options that mean the same thing are evaluated differently*, depending on how they are described. This is *not* a violation of the independence, or transitivity of preference. It is a violation of the (usually implicit) principle that the same decision problem should evoke the same preferences. This is the principle of *invariance*, and it is a normative principle of rational decision if anything is.

We see human psychology wandering further and further away from rational choice. Faced with risk aversion in the Allais problems, some theorists investigated ways of dropping the independence principle to get a theory that would accommodate observed behavior. Faced with Ellsberg problems, some theorists advocated dropping independence and some advocated weakening the requirement that preferences be totally ordered. Some think of their theories as purely descriptive—as psychology—but some think of them as normative. But to accommodate psychology, it appears that more has to go.*

Tversky and Kahneman conclude that the psychological, descriptive theory of choice and the logical, prescriptive theory of choice are two different enterprises. The prescriptive theory—the logic—is expected utility theory and an adequate descriptive theory must run a gamut of systematically observable deviations from it. This does not mean that people cannot *learn* the logic and implement it in situations where it is important to do so and there is enough time to think carefully. This is Kahneman's (2011) "thinking slow."

---

*In a large and growing experimental literature in psychology and behavioral economics, it appears that *almost all theories are systematically violated by some significant proportion of the population*. It also appears that there are different types in the population. Some violate one principle; some violate another. And there are even some expected utility maximizers.

We think that the proper perspective is that taken in the last chapter of *The Port Royal Logic*. The quote spans pages 367–68 and is from chapter XVI, "Of the Judgment which We Should Make Touching Future Events."):

> In order to judge of what we ought to do in order to obtain a good and to avoid an evil, it is necessary to consider not only the good and evil in themselves, but also the probability of their happening and not happening, and to regard geometrically the proportion which all these things have, taken together.
>
> These reflections may appear trifling, and they are so, indeed, if they stop here: but we may turn them to very important account; and the principal use which should be derived from them is that of making us more reasonable in our hopes and fears.[17]

Expected utility theory is a tool for improving the mind.

SUMMING UP

Humans often make mistakes in reasoning about chance, as they do in other realms of reasoning.

Some of these mistakes are systematically reproducible, not that everyone makes them, but many do. These include problems of Allais on risk aversion and of Ellsberg on preference for known chances over unknown chances. These are joined by a whole panoply of effects uncovered by Kahneman and Tversky. Systematic mistakes might be corrected by training in decision theory with important consequences—for instance, in medical decision making. Or systematic mistakes could be exploited for commercial or policy reasons. This leads to practical applications of behavioral economics.

Some of these mistakes—we believe that they are mistakes—violate one or another postulate of rational decision: independence or ordering, for instance. But others, emphasized by Kahneman and Tversky, show actual inconsistency. Individuals evaluate logically equivalent choices differently, depending on whether they are framed in terms of gains or losses.

The psychology of chance and the logic of chance come apart.

APPENDIX 1. ELLSBERG: ORDERING OR INDEPENDENCE?

We cannot have expected utility violation in the Ellsberg example if subjects obey both ordering and independence. But which is being violated? In order to answer this question, Ellsberg suggests we mark all balls in the first urn with I and all in the second with II and pour them into a single urn. This urn now contains 200 balls, either R_I or B_I, 50 R_{II} balls, and 50 B_{II} balls. We now ask about preferences among gambles on this combined urn. The following way of doing this was suggested to Ellsberg by Kenneth Arrow.

Consider the four gambles with payoffs in the following table. Suppose you are indifferent between I and IV, since both involve only risk— 100 balls pay off a and 100 pay off b. And suppose you are indifferent between II and III, since there still is no reason to think one of R_I, B_I is more probable, and both II and III have the same amount of uncertainty.

	R_I	B_I	R_{II}	B_{II}
I	a	a	b	b
II	a	b	a	b
III	b	a	b	a
IV	b	b	a	a

Ambiguity aversion would manifest itself in preference of I or IV over II or III. But we are supposing that you are indifferent between I and IV and between II and III. So if your preferences are an ordering, if you are ambiguity averse at all, you will prefer both I and IV to both II and III. In this case you violate independence.

You prefer I to II. By independence, your preferences should not depend on the cases where they give the same payoff but only on the other cases that are indicated in the following table in bold.

	R_I	B_I	R_{II}	B_{II}
I	*a*	*a*	*b*	*b*
II	*a*	*b*	*a*	*b*
III	*b*	*a*	*b*	*a*
IV	*b*	*b*	*a*	*a*

The same should be true for your preference of IV over III.

	R_I	B_I	R_{II}	B_{II}
I	*a*	*a*	*b*	*b*
II	*a*	*b*	*a*	*b*
III	*b*	*a*	*b*	*a*
IV	*b*	*b*	*a*	*a*

Everything depends on the payoffs for B_I and R_{II}. Given this, your preference for IV over III is inconsistent with your preference of I over II, as shown in the following table.

	B_I	R_{II}
I	*a*	*b*
II	*b*	*a*
III	*a*	*b*
IV	*b*	*a*

APPENDIX 2. DYNAMIC CONSISTENCY AND ALLAIS

Suppose you have Allais preferences that violate independence. Then you may have problems with dynamic consistency, as Howard Raiffa pointed out in 1968.[18]

Recall that then you prefer A over B in *problem 1,*

A: $1 billion on tickets 1–89   B: $1 billion with on tickets 1–89

+ $1 billion on tickets 90–100   + $0 on ticket 90
                                                      + $5 billion on tickets 91–100

and B over A in *problem 2:*

A: $0 billion on tickets 1–89	B: $0 billion with tickets 1–89
+ $1 billion on tickets 90–100	+ $0 on ticket 90
	+ $5 billion on tickets 91–100

This raises the question as to what you would prefer if you were told that the winner was not among tickets 1–89. That is, would you prefer

A': $ 1 Billion for sure	B': $0 with chance $\frac{1}{11}$
	$5 Billion with chance $\frac{10}{11}$.

Suppose you prefer A'. Then you will be in trouble with a version of problem 2. (If you prefer B' you will be in trouble with a version of problem 1.)

In problem 2 you prefer B over A, but if told that the winner was not among 1–89, you would reverse yourself and prefer A' over B'.

Suppose in problem 2, you own option A. You prefer option B, so you will presumably pay something, $e, to trade in A for B with a friendly broker. It is announced whether the winner is in 1–89. If not, you have lost $e. If so, the friendly broker now offers to trade back for some small price, $e'. Since you prefer A' over B', you accept.

You have lost $e + e'. Your broker has profited from your dynamic inconsistency by buying your asset from you and selling it back at a profit. [Should you be indifferent between A' and B', you will still trade back but won't pay. The broker can even pay you $e/2 to trade back and make a profit.] Any violation of the sure-thing principle is susceptible to this sort of example.

Jacob Bernoulli

CHAPTER 4

——

FREQUENCY

Early probabilists were well aware of the limitations of relying on intuitively equiprobable cases. The great seventeenth-century philosopher Leibniz had great hopes for the application of the new calculus of probability to practical affairs, to medicine, the law, and commerce. These concerns were shared by Jacob Bernoulli, and they carried on a lively correspondence.[1] They looked toward frequency evidence to inform probability judgments, as practical men had always done informally. Even today, if you ask a working scientist what probability $\frac{1}{3}$ means, you will usually get some version of the following: it means that in a long run of similar trials the event happens about $\frac{1}{3}$ of the time. In this chapter, we show the strengths and weaknesses of such an answer.

Leibniz and Bernoulli did not themselves identify probability with frequency. For them, probability was a form of rational degree of belief. What, then, was the formal connection between frequency and probability? Jacob Bernoulli succeeded in answering part of the question, with a version of our fourth great idea, the *law of large numbers*.

It establishes one of the most important connections between chance and frequency. The initial form, the weak law of large numbers, was proved by Bernoulli in his *Ars Conjectandi*.[2] It was subsequently strengthened to the *Strong Law of Large Numbers* by Borel and Cantelli, but this required a more powerful mathematical framework, as we shall see in the next chapter.

This great idea has a disreputable twin. It is the idea that chance *is* frequency—or close enough that it is not profitable to think about the matter more closely—the view held by many working scientists. The twin gains credibility by relation to his great brother. We start with the great idea in the seventeenth century, offer some cautions about the disreputable twin, and return to frequency in the twentieth century.

JACOB BERNOULLI AND THE WEAK LAW OF LARGE NUMBERS

Jacob Bernoulli proved the first law of large numbers. With arbitrarily high probability, the relative frequency of heads can be made to approximate the probability of heads as closely as you please by choosing a long enough series of trials.

Bernoulli aimed at a determination of the number of draws with replacement from an urn that would be required for the relative frequencies to be within specified bounds of the chances with a specified high probability ("Jacob Bernoulli on the Law of Large Numbers," tr. Oscar Sheynin. Online at www.sheynin.de/download/bernoulli.pdf):

> To make clear my desire by illustration, I suppose that without your knowledge three thousand white pebbles and two thousand black ones are hidden in an urn, and that, to determine [the ratio of] their numbers by experiment, you draw one pebble after another (but each time returning the drawn pebble before extracting the next one so that their number in the urn will not decrease), and note how many times is a white pebble drawn,

and how many times a black one. It is required to know whether you are able to do it so many times that it will become ten, a hundred, a thousand, etc., times more probable (i.e., become at last morally certain) that the number of the white and the black pebbles which you extracted will be in the same ratio, of 3 to 2, as the number of pebbles themselves, or cases, than in any other different ratio.

He immediately moves to a more careful statement; it is a question of the frequencies falling within bounds. And he remarks that if it were a question of the frequencies exactly equaling the chances, a long series of trials would just make things worse. At this point, frequencies and chances are clearly treated as two distinct and separate things.

Given the chances, the desired approximation interval, and the desired high probability of the frequency falling within the approximation interval, Bernoulli derived (with a tacit assumption of independence) an upper bound on the required number of trials. This was what he called his *golden theorem*. The law of large numbers follows.

As we saw, Bernoulli was interested in using the bound in empirical applications. But his bound was not very good and conjures up very large numbers of trials. *Ars Conjectandi* ends with an example. The chance is $\frac{3}{5}$, the desired interval for the relative frequency is between $\frac{29}{50}$ and $\frac{31}{50}$, and the desired probability that the frequency fall within that interval is $\frac{1000}{1001}$. Bernoulli's bound says that this is achieved if the number of trials is at least 25,550. We have datasets of this magnitude, but they were not available at Bernoulli's time.[3]

BERNOULLI'S SWINDLE AND FREQUENTISM

Bernoulli's motivation for his golden theorem was the determination of chance from empirical data. He was well aware that in many areas it was impossible to determine chances by counting symmetric cases ("Jacob Bernoulli on the Law of Large Numbers," tr. Oscar Sheynin. Online at www.sheynin.de/download/bernoulli.pdf):

. . . it would be an obvious folly to wish to find something out in this manner. Here, however, another way for attaining the desired is really opening for us. And, what we are not given to derive *a priori*, we at least can obtain *a posteriori*, that is, can extract it from a repeated observation of the results of similar examples. Because it should be assumed that each phenomenon can occur and not occur in the same number of cases in which, under similar circumstances, it was previously observed to happen and not to happen.

What does it mean to determine chances a posteriori from frequencies? The question is, given the data—the number of trials and the relative frequencies of success in those trials—what is the probability that the chances fall within a certain interval. It is evident that this is *not* the problem that Bernoulli solved. He solved an inference from chances to frequencies, not the inverse problem from frequencies to chances. The inverse problem had to wait for Thomas Bayes, whom we meet in chapter 6.

Yet Bernoulli somehow convinced himself that he had solved the inverse inference problem. How did he do so? It was by a vague argument using the concept of moral certainty. Bernoulli uses the term *moral certainty* to refer to a probability so close to 1 that for all intents and purposes, one may treat it as a certainty. Bernoulli's "general rule or axiom" number 9 is as follows:

> However, since complete certitude can only seldom be attained, necessity and custom desire that that, which is only morally certain, be considered as absolutely certain.

(He floats probability of $\frac{999}{1000}$ as a reasonable standard.)

Bernoulli argued that he had shown that with a large enough number of trials, it would be *morally certain* that relative frequency would be (approximately) equal to chance. But if frequency equals chance, then chance equals frequency. So, the argument goes, we have solved the problem of the inference from frequency to chance. This is Bernoulli's swindle. Try to make it precise and it falls apart. The conditional probabilities go in different directions, the desired intervals are of different quantities, and the desired probabilities that the quantity falls within the interval are different probabilities.

To be explicit, Bernoulli's conditional probabilities are probabilities about frequencies given chances, rather than probabilities about chances given frequencies.

They are probabilities that frequencies fall within a specified interval of frequencies given an exact value of chances, rather than probabilities that chances fall within a specified interval given an exact report of frequencies. Bernoulli gives the *chance* that frequencies fall within an interval given the *chance* on a single trial, rather than rational *degree of belief* that the *chances* fall within an interval given the frequencies.

The argument that the law of large numbers solves the inverse problem is a fallacy. But it is a rather slippery fallacy,* especially when stated informally. And it is a remarkably persistent fallacy, easy to swallow in the absence of rigorous thinking. We find it in the French mathematician and philosopher Cournot (1843),[4] who holds that small-probability events should be taken to be physically impossible. He also held that this principle (Fréchet named it Cournot's principle) is the one that connects probabilistic theories to the real world. It is taken, as in Bernoulli, as showing that we should identify probability with relative frequency in a large number of (independent? identically distributed?) trials.

This mantra was repeated in the twentieth century by very distinguished probability theorists, including Émile Borel, Paul Lévy, Andrey Markov, and Andrey Kolmogorov. We cannot help but wonder whether this was to some extent a strategy for brushing off philosophical interpretational problems, rather than a serious attempt to confront them.

Cournot's principle, taken literally, is absurd. Throw a dart at a target. The chance that it hits any specific point is very small. Then are we supposed to conclude that for any point, it is physically impossible that the dart hits that point? Later statements tend to try to get

---

* In the right setting, the conclusion of the argument may be approximately correct, but a fallacious argument to an approximately correct conclusion is still a fallacious argument. To assess the conclusion, we must go first to Bayes (chapter 6) and then to de Finetti (chapter 7).

around this by modifying the principle as saying that an event of very small probability, *singled out in advance*, is physically impossible. So you have to pick out a point in advance. Why should picking it out in advance make it physically impossible?

BERNOULLI'S SWINDLE AND HYPOTHESIS TESTING

By now, no theorist would be fooled by Bernoulli's swindle. But it seems to have an afterlife in popular thinking. Suppose a drug company runs randomized trials on a new drug. The drug is either effective or not. You would like to know the probability that it is effective given the data.

More people get better than get worse, but that might be just the way this sample came out. The drug company computes the probability that one would get the result in the data or better, given that the drug is ineffective, and gets a very small number.

To those who do not understand statistics, this is an invitation to Bernoulli's swindle. It is "morally impossible" to get this value if the drug is ineffective. Therefore, the drug is effective. The swindle is the same (except that moral impossibility comes at a rather low price). The untutored think that they are getting the probability of effectiveness given the data, while they are being given conditional probabilities going in the opposite direction. Nevertheless, the use of such *p*-values, is* a touchstone for publication in the biological and social sciences. This can have pernicious effects, a point that we will take up again in our chapter on Thomas Bayes.

The inventor of this methodology, Sir Ronald Fisher, was a brilliant statistician, and he was under no illusions as to what it

---

* Or was until recently. There is a movement toward reform. See, for instance, the recent statement by the American Statistical Association, Ronald L. Wasserstein and Nicole A. Lazar (2016): "The ASA's Statement on *p*-Values: Context, Process, and Purpose," *The American Statistician*, DOI:10.1080/00031305.2016.1154108. In a preamble to the official statement, the authors write: "Let's be clear. Nothing in the ASA statement is new. Statisticians and others have been sounding the alarm about these matters for decades, to little avail. We hoped that a statement from the world's largest professional association of statisticians would open a fresh discussion and draw renewed and vigorous attention to changing the practice of science with regards to the use of statistical inference."

showed. He said quite clearly that it did not give the probability of effectiveness given the data. He said what it did give you was a methodology and a story about why that is what you want. But it is not what you want, is it? You want the probability of effectiveness given the data.

HARD-CORE FREQUENTISM

There were frequentists, however, who saw Bernoulli's swindle as a fallacious argument and who tried to make do without it. This required some biting of the bullet, but they did not shrink from doing so. We discuss two such frequentists, John Venn and Richard von Mises.

VENN

In Victorian England there arose a strong movement that maintained that probability *was* relative frequency. Why not? Relative frequency is a proportion so—at least in the finite case—it obeys the laws of probability. Take the objects on my desktop now: a hat, two books, an umbrella, a pen, a case for eyeglasses. There are 6 objects in all. The relative frequency of hats is $\frac{1}{6}$, that of books is $\frac{2}{6}$, that of hats or books is $\frac{1}{2}$.

The Victorian frequentists—John Stuart Mill, Leslie Ellis, John Venn—claimed more than this. They claimed that relative frequency was the primary sense of probability and sometimes maintained that alternative senses were not important at all.

This was, to some extent, the English Empiricists against the Continental Rationalists. Mill was a proponent of the material view of logic, both inductive and deductive. Valid laws of inference got their warrant from their having always worked in the world of experience. Laws of arithmetic were empirical generalizations. From this point of view, it was natural for Mill to adopt a frequency view of probability. In the first edition of *A System of Logic* (1843), he ridicules Laplace's use of an ignorance prior. Here is a taste: "It would indeed require strong evidence to persuade any rational person that by a system of operations on numbers, our ignorance can be coined into science . . . ." But 3 years later, in the second edition, he changes his tune:

This view of the subject was taken in the first edition of the present work; but I have since become convinced that the theory of chances, as conceived by Laplace and by mathematicians generally, has not the fundamental fallacy which I had ascribed to it.[5]

He then goes on to endorse what is basically a subjective Bayesian position.

What happened? Mill's scientific colleagues set him straight. In particular, there was a notable exchange of letters with the astronomer Sir John Herschel, whom Mill had asked for advice in preparing the second edition.[6] Hershel replied that Mill had misunderstood Laplace and pointed out problems with Mill's naive frequentism. Mill replied that he was convinced, that he had gotten similar criticisms from others, and that he would thoroughly revise the discussion in the second edition—which he did.

If Mill had thoroughly understood Herschel's position and embraced it, he would have had to change quite a lot more of his philosophy, but he didn't. That is basically how it stood—a little Bayesian section in a non-Bayesian setting—for the rest of Mill's life.

It fell to John Venn to make a full-length exposition of the frequentist view in *The Logic of Chance* (1866).[7] Venn also spends a lot of time attacking the degree-of-belief probabilities of Laplace and of the prominent English Laplacian Augustus DeMorgan. But he also wrestles with the problem of the proper formulation of frequentism.

Probability is to be relative frequency in a suitable long series of events. Probability of heads is to be the relative frequency of heads in a long series of coin tosses. Probability of a male birth is the relative frequency of males in a long series of births. Probability of a human living past age 60 is just the relative frequency. It is clear at the outset that to talk of the probability of a single event is nonsense. There is no probability of this coin coming up heads when I flip it now, only a frequency of heads in a long series of trials. There is no probability of your next child being male or of Jane's living past age 60. There are only frequencies.

But to pin the theory down, we need to be more specific. What kind of series is required and how long should it be? Let us start with dice, which should be an easy case. If a die is tossed 10 times and only

10 times and comes up 2, 2, 6, 1, 4, 3, 6, 2, 5, 2, is the probability of getting a two $\frac{4}{10}$? That doesn't seem right. How about 20 times? What number is enough? Evidently only an infinite number of tosses will do, and this is the conclusion that Venn settles on. Probability is the limit of the relative frequency as the number of trials goes to infinity. We may already worry about the material view of logic. Does the real world contain an infinite number of such events?

The move to limiting relative frequency raises some mathematical issues, which we mention briefly and then put to the side. The first is that there are sequences such that the limiting relative frequency does not exist; as you move along longer and longer subsequences, the relative frequency fluctuates and never approaches a limit. The second is that the limiting relative frequency will typically depend not only on the trials, but also on their order. The third is that limiting relative frequencies may not add the way that finite relative frequencies do. Consider the integers in sequence. Each has a limiting relative frequency of 0, but taken together they have limiting relative frequency of 1. The probability of the infinite disjunction is not the infinite sum of the probabilities of the disjuncts.

The next question is what sort of events go into the sequence. Suppose that there are lots of dice, some symmetric and some oddly shaped, and lots of ways of throwing them. Should we just take any sequence of throws and use this to define probabilities? Evidently not. We want the same die, thrown in the same way. So let us take the same die, thrown in the same way. But, as Venn notes, if we keep throwing the same die again and again in the real world, it will alter its characteristics. It will become worn; edges and corners will round off. *It becomes a different die.* We want a sequence of trials in which all the relevant factors remain the same, so we can say that these are trials of the same kind. In the real world this appears impossible. So, even if the world can contain infinite sequences of events, the requirements that the sequence be of events that are appropriately similar appears in conflict with the requirement of an infinite sequence of trials. And if this is a problem with dice, how much more of a problem is it with sex or with lifespan—or with outcomes of a horserace or an election?

So, Venn finds himself pushed to a hypothetical relative frequency view. *The probability of this die coming up a 6 is the limiting relative frequency that it would have had if it were thrown an infinite number of times with no change in die, the tossing, or any of the relevant circumstances.* The material view of logic has been forced to define probability as a counterfactual! And the concept of relevant circumstances still appears to be a little vague. In a Newtonian world (as Arbuthnot had observed), if the die were always tossed the same way, it would always come up the same way. Must all probabilities then be either 0 or 1?

Venn thinks not. If a fair coin were tossed an infinite number of times (while remaining fair), the limiting relative frequency would presumably be $\frac{1}{2}$. Is that because there is a specific sequence of heads and tails, which is what we would get if this coin were tossed an infinite number of times and which has a limiting relative frequency of $\frac{1}{2}$, or does the counterfactual supposition deliver a class of sequences, all of which have limiting relative frequency $\frac{1}{2}$?

We seem to be dancing around the law of large numbers. Doesn't this answer our questions? The trials are relevantly similar if the probabilities are the same. If so, and the trials are independent, then with probability 1 the limiting relative frequency will exist and will equal the probability on a single trial. Can Venn use Bernoulli's swindle?

Venn cannot say any of this! Look at occurrences of "probability" in the foregoing statement. There is the probability on a single trial and the assumption that it doesn't change from trial to trial; there is the statement of independence of trials; there is the probability that relative frequency is equal to the single trial probability. *None of these probabilities are legitimate, according to Venn's frequentism.* He cannot state the law of large numbers!

It is to Venn's credit that he owned up to this. See where it leads him (Venn, *The Logic of Chance*):

> The reader who is familiar with Probability is of course acquainted with the celebrated theorem of James Bernoulli. This theorem, of which the examples just adduced are merely particular cases, is generally expressed somewhat as follows:—that in the long run all events will tend to occur with a frequency proportional to their objective probabilities. With the mathematical proof of this theorem we need not trouble ourselves, as it lies outside the province of this work; but indeed if there is any

value in the foregoing criticism, the basis on which the mathematics rest is faulty, owing to there being really nothing which we can with propriety call an objective probability.

If one might judge by the interpretation and uses to which this theorem is sometimes exposed, we should regard it as one of the last remaining relics of Realism, which after being banished elsewhere still manages to linger in the remote province of Probability.

In other words, the hard-headed, empirical Venn says that there are no chances in the world, only frequencies. And if large sections of probability theory go, then good riddance! But we have seen that in the last analysis, Venn's limiting relative frequencies are not of this world either. They live in a hypothetical world of infinite sequences of similar trials. *What connection, then, does this hypothetical world have to the actual world?*

There is another question that the frequency view leaves hanging. That is how probability can be—as Cicero said—a guide to life. How can probability as frequency inform degrees of belief and rational decision? Venn (*The Logic of Chance*) sees this as a question that must be addressed. His answer is roughly that as degrees of belief in a single event, we should take the corresponding relative frequency in a series of like events:

> . . . the different amounts of belief which we entertain upon different events, and which are recognized by various phrases in common use, have undoubtedly some meaning. But the greater part of their meaning, and certainly their only justification, are to be sought in the *series* of corresponding events to which they belong.

Our earlier question returns. How can we define *like events* without circularity? Intuitively we want them to have the same probability. Any single event belongs to any number of series, or reference classes. If you want to arrive at a judgmental probability about who will win the election, what is the appropriate series: all elections, all elections in your country, all elections with characteristics x, y, z, w, like the one in question? Soon you may be down to a series with one member. The relevant series then again can be nothing but a fictional infinite series.

Let us grant the idealization. Then Venn appears to claim that if one's degrees of belief agree with the relative frequencies, then an infinite series of fair bets would have a payoff in the limit that is fair, and that in general, expected value could be identified with limiting average value.

He has no right to claim this. Consider an idealized sequence of coin flips with limiting relative frequency of heads $=\frac{1}{2}$. Any bet at even odds is a fair bet. Consider an idealized agent—unlucky Joe—who bets on every toss and always loses. You might want to say that this cannot happen, but there is nothing in the frequency theory to preclude it. You might want to say that the probability of this happening is zero, but now you need some probabilities that are not available on the view to state your assumptions, do your proof, and state your conclusion.

Venn leaves us with many questions for the view that probability is frequency. Some are about the mathematical structure of the theory. Some are about the metaphysics of the theory, which is ultimately a theory of counterfactual or fictional series. Many are about the connection of the theory with real frequencies and real applications to decision.

If Venn's theory appears to be full of holes, it is to his credit that he saw most of them himself.

VON MISES

In his *Grundlagen der Wahrscheinlichkeitsrechnung* (1919), Richard von Mises set out to put the theory of probability on a sound mathematical basis. The challenge had been issued by David Hilbert in 1900 in his famous address to the international congress of mathematicians in Paris. He gave a list of 10 problems (later expanded to 23) to be solved by mathematics in the next century. Problem 6 was to give a mathematical treatment of the axioms of physics. Special emphasis was given to the role of probability in statistical physics:

The investigations on the foundations of geometry suggest the problem: *To treat in the same manner, by means of axioms, those*

*physical sciences in which mathematics plays an important part; in
the first rank are the theory of probabilities and mechanics.*

As to the axioms of the theory of probabilities, it seems to me
desirable that their logical investigation should be accompanied
by a rigorous and satisfactory development of the method of
mean values in mathematical physics, and in particular in the
kinetic theory of gases. (Hilbert 1900)

Von Mises was not responding to Venn but to Hilbert. The mathematics that he thought most relevant to the statistical physics of
Boltzmann* was a mathematics of frequencies. Nevertheless, the informal ideas of Venn were now to be made precise by a real mathematical mind. Von Mises interpreted probability as relative frequency in
a certain kind of infinite sequence, an idealization of the kind found
in statistical physics. This sort of a sequence—called a Kollektiv—was
to exhibit the intuitive ideas of local disorder and global order that
had been emphasized by Venn.

Global order meant the existence of limiting relative frequency.
Von Mises did not simply assume, like Venn, that limiting relative
frequencies would exist. It needed to be postulated as a defining at-
tribute of a Kollektiv. Local disorder was captured by a requirement
of randomness.

What is randomness? We are not talking about a random process but an extensional sense of randomness, such that a given infinite sequence of heads and tails (for instance) may be said to be
random or not. The answer is not obvious, and different approaches
might be taken. Von Mises' idea was that if a sequence was truly
random, the relative frequencies should be invariant under selection
of subsequences in some specified manner. For instance, the strictly
alternating sequence HTHTHTHTHT . . . is not random, because
in the subsequence of its odd members H has a relative frequency
of 1, and in the subsequence of its even members, it has a relative
frequency of 0.

Put in this crude way, the criterion is too powerful. If there is an
infinite sequence of heads and tails in which the relative frequencies

*Which we discuss in chapter 9.

are not 0 or 1, then that sequence contains an infinite subsequence of all heads and another of all tails. The way in which subsequences are selected must be somehow regimented. Von Mises considered *place-selection functions*, which map initial segments of the sequence to 0 (IN) or 1 (OUT). The leading idea here is *nonpredictability*. A further motivation for von Mises, which he thought came to the same thing, was the impossibility of a gambling system. If the relative frequency of heads in a Kollektiv of coin tosses was $\frac{1}{2}$, then a gambling strategy placing even-money bets should not be able to make money in the long run.

But which functions should be used as place-selection functions? If we allow any function, in the set-theoretical sense, then again for any sequence, there are functions that select out subsequences of all heads and of all tails. For a given sequence, map initial segments preceding heads onto IN and those preceding tails to OUT. That's your place-selection function, and it *is* a function in the set-theoretic sense, even if you don't like the way in which it was gotten. This objection was immediately pressed on von Mises, and as a result he relativized the account to some (unspecified) countable set of place-selection functions. Von Mises really had in mind a more restrictive, intensional notion of a function, but he did not have a means of making it precise.

The objection raises questions of the consistency of the theory. Do nontrivial Kollektivs even exist? Abraham Wald addressed the question in 1936 by proving that, for any finite or countably infinite collection of place-selection functions, there exists a sequence that is a Kollektiv according to them. There are, in fact, an uncountable infinity of such random sequences. Restrict your place-selection functions, and random sequences are common. Wald, like von Mises, was motivated by an intensional, constructive notion of a function.

This is where things stood for a long time, but it is a highly unsatisfactory resting point. What is needed is a natural class of place-selection functions that yields random sequences with nice properties. In 1940 Alonzo Church suggested an answer using the new theory of computability, which he himself had helped to create. Place-selection functions should be computable. Feed an initial segment of the sequence into a computer, and the machine will tell you whether the next element is in or out of the subsequence. This appears to be a very natural choice of place-selection functions. And since there are only countably many of these, by Wald's theorem, random sequences with respect to computable selection functions exist.

That would seem to be the end of the story, except for a complication noted by Jean Ville in 1939. Invariance of relative frequencies under computable place selection is satisfied by some sequences that do not have all the properties that we would want of a truly random sequence. In particular, Ville showed that relative to any countable class of place selection functions, there is a Kollektiv in which relative frequency of H is $\frac{1}{2}$, but for all but a finite number of initial segments, the relative frequency is not less than $\frac{1}{2}$. The limiting relative frequency is approached from above. We don't have fluctuations above and below, as one would expect. This allows the very possibility of a gambling system that von Mises initially wished to rule out. In fact, the very sequences that Wald constructed to prove the consistency of the conception of a Kollektiv have just this character.

Invariance of relative frequency under place selection does not guarantee impossibility of a gambling system, as von Mises thought. This leaves the very interesting question of a satisfactory definition of a random sequence still open. We will return to this later, in chapter 8.

IDEALIZATION REVISITED

Von Mises, like Venn, was sensitive to the finite and imperfect nature of sequences of real events. He provided a thoughtful discussion of the relation between Kollektivs and the real world. Von Mises held that Kollektivs are no different in status from other mathematical idealizations used in science (Richard von Mises, *Probability, Statistics and Truth*, 2d rev. English ed., prepared by Hilda Geiringer (New York: Dover, 1981)):

> Although a straight line in geometry is defined completely abstractly through its axioms and the sphere through the property (which in a concrete case can never be verified) that all its points have the same distance from the same fixed point, nevertheless the geometrically derived relations between the

straight line and the sphere are applicable in the realm of build-
ings, and so on. Of course precision and infallibility are lost in
this process.

It is to be the same way with the Kollektiv.

> To understand the "scientific" theory of probability, that is to be
> the subject of these lectures, one has always to bear in mind the
> complete analogy with geometry.

The relation of mathematical scientific theories to the world is a
deep question, and it is not obvious that it has a univocal answer. It
might have different answers in different applications. In particular,
is the finding that a physical line segment is approximately straight
from physical measurements completely analogous to finding that a
sequence is approximately random, or that its limiting relative frequency is approximately $\frac{1}{2}$ from inspection of a finite initial segment?
In the former case, it seems that our imprecise measurements can at
least establish bounds, which is not true in the latter.

In any case, is there a *methodology* for connecting the mathematics
to reality, or must the discussion remain at the informal level? This is
a question that alternative views also need to address. Finally, if idealized mathematics is permissible in general, why should the theory of
probability be restricted to Kollektivs? Could not the use of a more
general theory of probability have an analogous status?

Could we not just have *objective chances*, which apply to single
cases as well as wider classes of events, as theoretical scientific entities?
There is a chance that this coin comes up heads on this toss. There are
chances of sequences of outcomes in sequences of trials. These chances
may or may not be such that the trials are independent. Frequency
evidence could still be brought to bear on the validity of a chance
model. This is what is now called a *propensity* view of probabilities.
This seems to be the view of Fréchet,[8] who views chances as physical
quantities with objective existence in the world.

The proper mathematical framework within which to formulate
chance models would then be the measure-theoretic framework that
is the topic of chapter 5 rather than the apparatus of Kollektivs of von
Mises.

SUMMING UP

What is the relation of frequency to probability? Bernoulli (and then Borel) gave part of an answer in terms of laws of large numbers. Probabilities of single cases are essential in stating and proving the laws. The frequentist views of Venn and von Mises seem to help themselves to the conclusions of these theorems *by definition*.

Von Mises postulates kinds of infinite sequences that would typically (with probability 1) be produced by independent and identically distributed trials. After von Mises, Wald, and Church, the notion of an objectively random sequence began to emerge with some mathematical precision, but as a general account of probability, it suffers shortcomings that were apparent to Venn.

Frequentism radically restricts the range of probability theory, and it must be judged inadequate as a general account of probability. But a deep question has been uncovered in the development of the frequentist view. *What is the nature of a random sequence?*

Furthermore, problems of interpreting frequentism have raised this question: *What is the relation of idealization to reality?* In frequentism, it arose because the development of a theory demanded more than actual frequencies in the world. It required limiting relative frequencies in idealized infinite sequences. But any theory in which there is an idealized probability, frequentist or not, must face this question. Is there any principled answer? We will return to each of these questions in subsequent chapters.

Andrei Kolmogorov

CHAPTER 5

MATHEMATICS

Our fifth great idea is the formulation of probability theory as part of mathematics within the modern theory of measure and integral. This was a project that became salient with the development of measure theory in the early twentieth century. It was completed by Andrei Kolmogorov in a monograph published in 1933.[1] Before moving to Kolmogorov's contribution, we discuss the background, treat the mathematization of finite problems (there are still things to think about), and take a first pass at the infinite version of the law of large numbers.

BETWEEN MATHEMATICS AND REALITY

The connections between mathematics and probability (or indeed any area of applied science) is a lively, controversial subject even today. To appreciate the issue, consider a typical probability problem: Let $X_1, X_2, \ldots$ be independent random variables. What is the chance of . . . ? A natural question: What is the definition of random variable? Classically, and in many of today's textbooks, you see definitions such as, a random variable is the observed value of a random quantity. What on earth does that mean? How can any sort of theory be built on such vagueness?

Beginning students in any science may face similar difficulties. Consider basic mechanics. For most of us length and time are OK, and so velocity and acceleration make sense (maybe even mass). But what is force? It's not so simple, and a perusal of beginning physics texts is less than satisfying. (You'll find force defined by $F = ma$, with the assertion that this works in the world.)

It is natural to ask for precise definitions, both for intellectual clarity and to avoid mistakes. There are numerous "physics proofs" of mathematical theorems.[2] Are they OK? The great geometers of the late nineteenth century developed marvelous facts using geometric intuition. You know, things such as, consider two 4-dimensional manifolds in 5-dimensional space intersecting transversely . . . . The problem was that some of the "facts" and "theorems" turned out to be wrong. It took a century of modern algebraic geometry to straighten things out. Similarly, marvelous feats of probabilistic intuition have led both to great theorems and to real errors.

FINITE PROBABILITY

The early theory of probability as counting (our chapter 1) seems perfectly clear. Let us start by shining a light there. When we begin a first course in probability, we start with some version of the following: probability is part of mathematics used to model random phenomena. For example, a question is posed in English, such as, Flip a fair coin 4 times. What is the chance of 2 heads? In mathematics we introduce a sample space, χ, which is just the set of all possible outcomes:

$$\chi = \{0000, 0001, 0010, 0011, 0100, 0101, 0110, 0111,$$

$$1000, 1001, 1010, 1011, 1100, 1101, 1110, 1111\},$$

with the symbols 1 and 0 standing for heads and tails, respectively.

We then introduce a probability distribution over *events*, which are sets of members of the sample space. Members of the sample space will be called *points*. For instance, the event of "two heads" is the set of all points with 1 occurring twice:

$$A = \{0011, 0101, 0110, 1001, 1010, 1100\}.$$

We make the points in the sample space equiprobable: $P(x) = \frac{1}{16}$ for each possible outcome, and define $P(A)$—the probability of A—as the sum of the $P(x)$ for all x in A.

$$\text{Here } P(A) = \tfrac{6}{16} = \tfrac{3}{8}.$$

All this makes a clear piece of mathematics. (However, don't lose sight of a basic philosophical question: what does mathematics have to do with flipping real coins? We will return to this question throughout this chapter.) Almost all classical probability can be translated this way:

There is a finite set, χ, and a positive probability, $P(x)$ for all points in χ (with these numbers summing to 1). Then, given a set of points A in χ, the basic problem of probability is to compute or approximate $P(A)$, which is defined as the sum of $P(x)$ for all x in A.

INFINITY I

Toward the end of the nineteenth century a fresh challenge arose. Bernoulli's law of large numbers showed that if a fair coin was flipped n times, the proportion of heads would be close to $\frac{1}{2}$, with arbitrarily high probability for an appropriately large number of tosses.

All this can be carefully understood using the model of coin tossing, with the sample space χ being the set of 2^n binary sequences of length n, and $P(x) = \frac{1}{2^n}$. Émile Borel asked a different, more demanding, kind of question: what is the chance that the proportion of heads gets (arbitrarily) close to $\frac{1}{2}$ and stays there forever?

The answer is that the chance is 1. This is the strong law of large numbers.

Borel asked a question about *infinitely* many tosses. The chance of any particular infinite sequence of outcomes is 0. But something has to happen. Are we to add up infinitely many 0s and get something positive? Actually, yes! How this was to work was not obvious. This new development called for a fresh mathematization.

Borel introduced a seemingly very different sample space χ and probability P. He took χ to be [0,1], the set of real numbers between 0 and 1. For the real numbers in an interval A, he took $P(A)$ to be the length of A. This was then extended to more-complicated sets by adding up lengths of intervals, as set forth in our appendix. This seems much further removed from tossing real coins. But we will see that is not as crazy as all that.

Write a point in binary: 0.01001100. . . . The binary digits stand in for our coin tosses. Consider the set of all points that start with 1: 1*** . . . , where the *s can be either 0 or 1. These are the points in the right half of [0,1]:

$$0 \underline{\hspace{4cm}} \tfrac{1}{2} \underline{\hspace{4cm}} 1.$$

The length of this interval, $\tfrac{1}{2}$, is equated with the chance that the first flip comes up heads. Similarly, all points that start out with 0: 0*** . . . make up the left half of the interval, which also has length (equals probability) $\tfrac{1}{2}$. The points with a 1 in position 2:

*1*** . . . make up the second and fourth quarters of [0,1]:

$$0 \underline{\hspace{2cm}} \tfrac{1}{4} \underline{\hspace{2cm}} \tfrac{1}{2} \underline{\hspace{2cm}} \tfrac{3}{4} \underline{\hspace{2cm}} 1.$$

It thus also has probability $\tfrac{1}{2}$.

More is true. The probability of a 1 in both the first and second position equals the length of the fourth quarter of the interval, which equals $\tfrac{1}{4}$. Similarly, the probability that the first n positions have any fixed pattern is $\tfrac{1}{2^n}$. This abstract model contains *all* our finite coin-tossing models in one neat package.

It contains much more; we can now speak of infinite events. This can lead to a seeming mess. Consider the relatively clear English sentence:

The probability of heads gets arbitrarily close to $\tfrac{1}{2}$ and stays there.

What is the corresponding set of points in [0,1]? Just to show off, here it is (drum roll please):

$$(*) \bigcap_{k=1}^{\infty} \bigcup_{n=1}^{\infty} \bigcap_{m \geq n}^{\infty} \left\{ x : \left| \frac{x_1 + \cdots + x_m}{m} - \frac{1}{2} \right| < \frac{1}{k} \right\}.$$

This is a pretty complicated set. But the early workers (Borel, Cantelli, Faber, Hausdorff) showed that it has length/probability equal to 1. This is the *strong law of large numbers*:

With probability 1, the averages get arbitrarily close to $\frac{1}{2}$ and stay there forever.

Take a look at the horrible expression in *. The inner set, the expression here,

$$\left\{ x : \left| \frac{x_1 + \cdots + x_m}{m} - \frac{1}{2} \right| < \frac{1}{k} \right\},$$

is the set of all points in [0, 1] such that the average of the first m coordinates is within $1/k$ of $\frac{1}{2}$. The innermost intersection is the set of all points for which this is true for all m greater than or equal to n. The middle union says that one of these happens for some trial n. (That is, for at least one n, past n averages are within $1/k$ of $\frac{1}{2}$.) The outer intersection says that this happens for all k (so the averages are arbitrarily close to $\frac{1}{2}$).

Assigning length to such esoteric sets was possible because of the work of Borel, Lebesgue, and others. It turns out that not every set can be assigned a length (see our appendix.) At the end, Borel and his coworkers had a rigorous mathematical model with real similarities to coin tossing, with which they could use new mathematics to compute answers to interesting questions.

There was more to do.

Consider the following recollection by a great probabilist, Mark Kac:[3]

When I came to Lwow as a student in 1931, I had never heard of probability theory. A few Polish mathematicians had contributed a number of isolated papers on the subject . . . but for all intents and purposes, the subject did not exist. . . .

It must have been in 1933 or 1934 that I came across A. A. Markov's *Wahrscheinlichkeitsrechnung.* This book, a 1912 translation of the original Russian, made a tremendous impression on me even though I could not really follow it. It was not that I found the technicalities of the moment a problem and related analytic subtleties difficult to absorb. This part was relatively easy. What I could not make out was what were the "random quantities" X_1, X_2, . . . to which this elaborate and fascinating theory was supposedly applicable.

Kac knew all about measure theory, but in the 1930s people still hadn't internalized the connection. He records his amazement on reading Kolmogorov's unification.

Kac wasn't alone in wondering how to mathematize probability.

HILBERT'S SIXTH PROBLEM

The great mathematician David Hilbert set out a list of 23 problems at the start of the twentieth century. His sixth problem asks (David Hilbert, "Mathematical Problems," *Bulletin of the American Mathematical Society* 8, no. 10 (1902): 437–79):

> The investigations on the foundations of geometry suggest the problem: *To treat in the same manner, by means of axioms, those physical sciences in which mathematics plays an important part; in the first rank are the theory of probabilities and mechanics.*
>
> As to the axioms of the theory of probabilities, it seems to me desirable that their logical investigation should be accompanied by a rigorous and satisfactory development of the method of mean values in mathematical physics, and in particular in the kinetic theory of gases.

Hilbert had in mind Boltzmann's theory of gases, which we will revisit in chapter 9. An ideal Boltzmann gas consists of a container of hard spheres, which move around, interacting with one another and with the walls of the container only by elastic collisions. Observed macroscopic variables are the result of the motion with, for instance, observed pressure on the walls being the result of collisions of the spheres with the walls. Assuming some prior probability on microstates of the system (specifying velocity and momentum of each sphere), can we show that with high probability low-entropy states will evolve to high-entropy states?

Although Boltzmann used probabilistic arguments, they were on an ad hoc and semiformal basis. Application of probability theory to this problem requires a framework within which a rigorous theory of stochastic processes can be developed. Two tools that were lacking were (1) a theory of conditional expectation and conditional probability that applied to continuous variables and (2) an account of the relation of a stochastic process in infinite time (discrete or continuous) to the finite-dimensional distributions that it comprises. These would both be supplied by Kolmogorov.

WHAT DID KOLMOGOROV DO?

Andrey Kolmogorov, one of great mathematicians of the twentieth century, worked in almost all areas of mathematics. His 1933 monograph, *Grundbegriffe der Wahrscheinlichkeitsrechnung* (Foundations of the Theory of Probability), does three main things:

1. He puts the theory of probability on a clear mathematical foundation.
2. He properly formalizes conditional probability.
3. He proves the Kolmogorov extension theorem.

In addition, the monograph is peppered with smaller beautiful gems (Kolmogorov's 0–1 law) and presents a unified treatment of his earlier work on limit laws in probability for general random variables. The following synopsis gives brief descriptions of these highlights. It is remarkable that Kolmogorov's formulation is still the absolute standard for all flavors of probability: subjective, objective, and everything in between.

PROBABILITY AS MATHEMATICS

In his preface, Kolmogorov writes (A. Kolmogorov, (1933) "Grundbegriffe der Wahrscheinlichkeitrechnung," *Ergebnisse der Mathematik* (1933); English translation: *Foundations of Probability* (New York: Chelsea, 1950))

The author set himself the task of putting in their natural place, among the general notions of modern mathematics, the basic concepts of probability theory concepts which until recently were considered to be quite peculiar.

The treatment is abstract and axiomatic. This is not a book about interpretation; it is about the mathematical structure of probability theory.* This is emphasized from the onset, in words that echo those of Hilbert:

The theory of probability, as a mathematical discipline, can and should be developed from axioms in the same way as Geometry and Algebra. This means that after we have defined the elements to be studied and their basic relations, and have stated the axioms by which these relations are to be governed, all further exposition must be based exclusively on these axioms, independent of the usual concrete meaning of these elements and their relations.

The mathematical object of study for probability is introduced in a way familiar to modern students of the subject. It is a triple:

$$\langle X, \mathcal{F}, P \rangle,$$

where X is some set of objects, $\mathcal{F}$ is a set of subsets of X, and P is a real-valued function defined on $\mathcal{F}$. The members of $\mathcal{F}$ are the things that have probabilities; P assigns each of them a numerical probability.

These probability spaces are governed by axioms:

$\mathcal{F}$ is closed under countably infinite Boolean operations, union, negation, and intersection.[†]

P maps members of $\mathcal{F}$ to nonnegative reals, such that $P(X) = 1$ and P is countably additive.[4, 5]

---

*There is, to be sure, a perfunctory section on the relation to experimental data, in which Kolmogorov tips his hat to von Mises, but it seems that he attaches no great importance to this. He suggests that the reader might well skip it.

[†]The countably infinite operations allow quite complicated sets to be constructed. If we spin an idealized wheel of fortune of unit circumference, what is the probability that the uppermost point comes out to be in the set of rational numbers or in one of the more complicated sets constructed by Cantor?

Here is a very simple and clean framework. Any geometrical considerations have been abstracted away; X is just a set. If, in a particular application, it has relevant structure, then that is addressed in the application rather than the general theory. The appearance of $\mathcal{F}$ makes room for sets that do not have any probability at all; they are just not in $\mathcal{F}$. In some applications, all subsets of X may be in $\mathcal{F}$, but in other applications some subsets of X may not be measurable—probability just does not apply. (For more on nonmeasurable sets, see appendix 2 in this chapter.)

In this framework, it is easy to give a precise definition of a real-valued *random variable*, the notion that Kac found mysterious in early expositions of probability theory. It is a measurable function from the basic set X to the real numbers. One can think of the mass of a widget coming off an assembly line or the lifespan of a human. One would want each interval, for example, mass between 1 and 2 grams, to have a probability—and more generally for a measurable set of values of the variable to have a probability. This is just what it is to say that *the function is measurable*; the inverse images of its measurable sets are members of $\mathcal{F}$, the sets that have probabilities. Random quantities are no longer mysterious objects. They are just measurable functions. The expectation of a random variable is an integral.

CONDITIONAL PROBABILITY AS A RANDOM VARIABLE

Conditional probability is now generalized to *conditional probability as a special kind of random variable*. The traditional conditional probability, $P(B|A)$, is defined as $P(A \text{ and } B)/P(A)$ when $P(A) > 0$ and undefined otherwise. For a simple example, we consider a finite partition. In fact we partition our population into just two classes, men and women. The probability of eventually dying of heart disease has one value for men, $P(\text{heart}|\text{men})$, and another for women, $P(\text{heart}|\text{women})$. We can define the probability of dying of heart disease *conditional on sex*, $P(\text{heart}\|\text{sex})$, as the random variable that takes the first value for men and the second value for women.

Conditional expectation as a random variable is defined in the same way. For example, life expectancy at birth *conditional on sex* is a random variable that takes one value for each male (life expectancy of males) and another value for each female. The treatment of conditional probability and conditional expectation as random variables is straightforward and intuitive in these simple cases, but when pursued in generality it has very important consequences.

With continuous random variables, we may often want to condition on a probability 0 event. Conditional probability as a random variable allows us to do this. The important idea is to focus on the role of conditional probability in calculating the total probability of a set by adding up the probabilities of its parts.

This idea is easy to see in the context of a finite partition $o_1, o_2, \ldots,$ o_n of our probability space. But now we consider the case in which some of the outcomes have probability zero. (We could have a special category for males with a theoretically possible chromosome abnormality that is never seen.) We can define a function, f, which for every point in o_i takes as its value the traditional conditional probability, $P(B|o_i)$, if defined. If it is not defined because the experimental result has zero probability, we can just put in any probability value! This function is still called *probability conditional on the partition* and is denoted $P(B\|\mathbf{O})$. (Different arbitrary choices for experimental outcomes with zero probability give different functions that are said to be different *versions* of $P(B\|\mathbf{O})$.) Now what is important is that all the different versions give the same answer to some questions that we want to ask.

We want to determine the total probability of B as the sum of the probabilities of members of the partition multiplied by probability of B conditional on the member. That is to say that the probability of B is just the expectation of $P(B\|\mathbf{O})$. The arbitrary differences of different versions wash out because they are multiplied by zero probabilities in determining the total probabilities.

That is all perfectly simple. Moving to the general case requires a little heavier artillery. *An experiment is a sigma algebra* of sets in $\mathbf{F}$ (closed under countable Boolean operations) rather than just a partition. An experimental result can be thought of as telling us the elements of the algebra in which the actual point lies. Conditional probability as a random variable is now relative to an algebra rather than just a partition. This algebra includes complements and countably infinite conjunctions and disjunctions of its members—it is called a *sigma algebra*. The importance of conditional probability, as before, lies in its expectation providing total probability. If *f* is a version of probability of *B* conditional on the sigma algebra, **O**, then for any member of the algebra, *A*, the probability of *A* and *B* can be gotten by integrating this function:

$$P(A \text{ and } B) = \int_A f \, dP.$$

Different versions of conditional probability differ on a set of probability 0, but as before, the differences wash out in the integral.[6]

This construction makes explicit the dependence of conditional probability as a random variable on the underlying sigma algebra. This was sometimes overlooked in earlier discussions of conditioning on an event of probability zero, especially in geometrical contexts where there is a tendency to rely on geometrical intuition. As an illustration, Kolmogorov analyzes the *Borel paradox* and shows that conflicting intuitions are merely the reflection of the choice of different sigma algebras.

FROM FINITE TO INFINITE DIMENSIONS

Kolmogorov now shows how an infinite-dimensional stochastic process can be built up from a consistent family of finite-dimensional probability spaces. This is the *Kolmogorov extension theorem*.

For example, consider repeated trials of a spin of a wheel of fortune with unit circumference. For a single trial we have a measurable space $S^1 = \langle E^1, F^1 \rangle$, with the points in E^1 being real numbers in $[0, 1)$ and the events in F^1 being the Borel sets.[7] Two trials are represented in the product space $S^2 = S^1 \times S^1$ and likewise for *n* trials and for a countable sequence of trials. Probability measures on S^1 and S^2 are *consistent* if the probabilities on S^1 gotten by restricting the probability on S^2 to the sets that are the natural counterparts of those in S^1 are the same. The same holds for S^n and S^m, $m > n$.

If we start from a probability on the infinite-dimensional space S^∞, then it obviously determines a consistent family of probabilities in the respective finite subspaces. The probability of a set of sequences of length n in S^n is just the probability of the set of all possible ways to continue them in S^∞. The Kolmogorov extension theorem shows that you can go the other way. A consistent family of measures in the finite-dimensional spaces induces a unique measure on the infinite-dimensional space. Time need not be discrete, as in our example, but may be continuous, so it is applicable to continuous time stochastic processes.

Kolmogorov used his generalized concept of conditional probability to give a rigorous definition of Markov processes. And he proved the definitive form of the strong law of large numbers for general random variables: the average of a sequence of independent random variables converges if and only if it has finite mean values. His measure-theoretic framework was soon adopted as the standard setting for the study of stochastic processes.[8]

BACK BETWEEN MATHEMATICS AND REALITY

After Kolmogorov, probability became a healthy part of mathematics, developed as much for its own internal logic and beauty as for applications. This leaves the following question: when are the calculations and theorems of probability applicable to real world phenomona? This is not just philosophical nitpicking.

One growth area for applications has been financial mathematics. If the logarithms of stock prices indeed fluctuate like Brownian motion, then the findings about Brownian motion can be used to make predictions and set prices. Based on such assumptions, mathematicians and economists derived pricing formulas, such as the Black-Scholes formula for options pricing (for which Myron Scholes and Robert Merton were awarded the 1997 Nobel prize in economics.) Alas, it was a castle built on sand. Brownian motion—a version of the bell-shaped curve—is a useful approximation "in the middle." But it was being used too far out in the tails. Those fall off like $e^{-x^2/2}$. In practice, far too many "rare events" actually occurred, and real financial markets were left in meltdown. Spirited accounts can be found in Nassim Taleb's 2007[9] *The Black Swan* and Andrew Lo and Achie MacKilnay's 1999 *A Non-Random Walk Down Wall Street*.[10] David Freedman's careful studies of the misuse of probability models in the social sciences[11] give dozens of further cautionary examples.

These problems are not unique to probability; they appear throughout applied mathematics. But they should remind us that the more idealized and abstract the mathematics, the more we need to be careful about applications to the real world.

INTEGERS PICKED AT RANDOM?
A MATHEMATICAL ASIDE

Here is a standalone study of the tension between intuition and rigor. The problem discussed is still not settled, so the reader is welcome to weigh in. Throughout, we work with the set of natural numbers, $N = \{0, 1, 2, 3, \ldots\}$. What could be simpler?

Consider this question: pick a number at random. What is the probability that it is even?

It seems to most people that half the numbers are even, so the answer is $\frac{1}{2}$. This is actually an age-old question. The Medieval philosopher Oresme (commenting on Cicero's account of an example of Greek philosopher Carneades) said

> The number of stars is even, the number of stars is odd. . . . We have doubts . . . . The number of stars is a cube. We say that it is possible, but not credible or probable or likely, since such numbers are much fewer than others. . . . The number of stars is not a cube.[12]

By similar reasoning, a random number divided by 3 would show remainders of 0, 1, and 2, each $\frac{1}{3}$ of the time, and so on for any fixed number ($\frac{1}{5}$ of the numbers being divisible by 5).

Does this kind of talk cohere? How can we make sense of it? One way is to work with finite sets and their limits. Let A be a set of numbers (for example, the even numbers, or the primes, or the squares). Look at

$$\frac{\text{no. \{numbers in } A \text{ less than } n\}}{n}.$$

Thus, if A is the even numbers, $A = \{2, 4, 6, 8, \ldots\}$ and $n = 5$, this ratio is

$$\frac{\text{no. } \{0, 2, 4\}}{5} = \frac{3}{5}.$$

If $n = 10$,

$$\frac{\text{no. } \{0, 2, 4, 6, 8\}}{10} = \frac{5}{10} = \frac{1}{2}.$$

If $n = 100$, the ratio is $\frac{50}{100}$; if $n = 101$, the ratio is $\frac{50}{101}$. For large n, the ratio approaches $\frac{1}{2}$.

Let us say that A has a *density* θ if this ratio tends towards the limit, θ. The odd numbers have a density $\frac{1}{2}$ and the multiples of 5 have a density $\frac{1}{5}$. What about taking A to be the set of squares:

$$A = \{0, 1, 4, 9, 16, 25, \ldots\}.$$

The proportion of squares less than n is about $(\sqrt{n})$. This tends to 0 when n is large, in accordance with intuition. Similarly, the set of primes has density 0, so we would like to say that the chance of a random number being prime is 0!

Using density as a careful version of probability in these settings runs into problems:

– Not all sets have a density.
– Density is not countably additive.

Here are examples for each of these problems.

A: Set Without Density

Look at the numbers on the front page of today's *New York Times*. What proportion of these begin with a 1? The suprising answer is about 30%. To understand, let

$$A = \{1, 10, 11, 12, \ldots, 19, 100, 101, \ldots, 199, \ldots\}$$

be the set of numbers that begin with 1. If this set had a density of about 0.3, all would make sense. Alas, A does not have a density at all! When $n = 9$, only one of the numbers up to 9 are in A; when $n = 20$, $\frac{11}{20}$ of the numbers are in A; when $n = 200$, the proportion is back to about $\frac{1}{2}$. It continues to oscillate forever.

Yet, there is a sense in which A has 0.3 (actually $0.301 \cdots = \log_2 10$) as a limit. This will be discussed later in this chapter. There is a charming book-length discussion of this first-digit problem by Berger and Hill.[13] This even has some applications; naturally occurring sets of numbers begin with a 1 about 0.3 of the time. If people are faking data, they tend to make the proportion closer to $\frac{1}{9}$. The Internal Revenue Service and political scientists who study faked elections use these observations as a test.

B: Failure of Additivity

Suppose we try to make sense of things by defining the probability of A to be its limiting density *if* this density exists. After all, not all sets of points in the plane have their area defined, but length on a line and area in a plane work perfectly well when restricted to the class of measurable sets. A new problem arises. A single point, for instance {5}, has density zero, but this is true for each single point, and yet the set of all points {0, 1, 2, 3, . . .} has density 1. There are variants of this problem. There are sets A and B that have densities such that their union lacks a density.[14]

There are various fixes for these problems. Bruno de Finetti drops the requirement of countable additivity. There *are* finitely additive measures, defined for all subsets of numbers, which agree exactly with density, whenever density exists. Here are two points to consider. First, there are many different extensions of density, and one must just choose. For de Finetti, probability is an expression of subjective belief, so this might actually be viewed as a plus. Put positively, if you have assigned probabilities to a collection of sets $A_1, A_2, A_3, \ldots$ (and their complements and intersections) in a coherent manner, given a further set, A, there is always a coherent extension of your probabilities that allows a choice of $P(A)$. A second drawback is that actually specifying an extension of density to all subsets of numbers needs nonconstructive set theoretic axioms. This doesn't bother de Finetti much. He is really a finitist, and he is not interested in leaping to a probability on all subsets. His theory says that you can always keep extending, one step at a time.

Another way to go is to stick with ordinary probability, weighting the earlier numbers more heavily. We are more likely to encounter smaller numbers that larger ones.

This could be done in various ways. If we gave probability $\frac{1}{2}$ to 0, probability $\frac{1}{4}$ to 1, and so forth, the probabilities would add up to 1, but our nice results about densities would be trashed. But we can give the numbers any series of weights that converge to some finite number and then divide by that number to convert them to probabilities. A series of weights that converges slowly can deliver results that approximate our densities as closely as you please.

Suppose we let our weights, for fixed s larger than 1, be

$$P_s(j) = \left(\frac{1}{j^s}\right)\left(\frac{1}{\text{normalizer}}\right) \text{ for } j = 1, 2, 3, \dots .$$

The normalizing constant, by which we divide to make things add up to 1, depends on s and is given by the celebrated Riemann zeta function. Here are approximate values of the normalizing constant for various values of s:

> If $s = 1.1$, normalizer is about 10.6.
>
> If $s = 1.01$, normalizer is about 100.6.
>
> If $s = 1.001$, normalizer is about 1000.6.
>
> If $s = 1.00001$, normalizer is about 100000.6.

When s is close to 1,

$$P_s(\text{even number}) = \tfrac{1}{2^s} \approx \tfrac{1}{2},$$

$$P_s(\text{multiples of } 5) = \tfrac{1}{5^s} \approx \tfrac{1}{5}.$$

What is more, our two previous difficulties are avoided: (1) This assignment is countably additive, and (2) all sets of natural numbers can be assigned a definite probability. Finally, if a density exists, P_s agrees (approximately) with the density.

Now (!!!), for the first digit set, as s tends to 1,

$$P_s(\text{numbers that begin with } 1) \approx \log_{10}(2) = 0.301. \ldots$$

One drawback is that we have to choose an s, but as long as s is close to 1 (e.g., 1.00001) the choice of s hardly matters.

All this is accomplished within Kolmogorov's framework.[15]

KOLMOGOROV'S VIEW OF THE INFINITE IN PROBABILITY SPACES

Kolmogorov makes some interesting remarks about the relation of the finite to the infinite. Preliminary to these remarks he proves the Carathéodory *extension theorem*:

For any field of sets, **F**, there is a unique smallest Borel field (sigma field) that contains it. Call this the Borel extension of the field, $B\mathbf{F}$. Then it is proved that

A completely additive probability on a field, **F**, can be extended to a completely additive probability on its Borel extension, $B\mathbf{F}$ and this can be done in only one way.

(This theorem is then used in the proof of Kolmogorov's extension theorem.) Following the proof, Kolmogorov inserts a brief discussion of the infinite elements introduced in this way (A. Kolmogorov, "Grundbegriffe der Wahrscheinlichkeitsrechnung," *Ergebnisse der Mathematik* (1933). English translation: *Foundations of Probability* (New York: Chelsea, 1950)):

Even if the sets (events) of **F** can be interpreted as actual and (perhaps only approximately) observable events, it does not, of course, follow from this that the sets of the extended field $B\mathbf{F}$ reasonably admit of such an interpretation.

Thus there is the possibility that while a field of probability $(\mathbf{F}, P)$ may be regarded as the image (idealized however) of actual random events, the extended field of probability $(B\mathbf{F}, P)$ will still remain merely a mathematical structure.

Thus the sets of $B\mathbf{F}$ are generally merely ideal events to which nothing corresponds in the outside world. As an illustration we are asked to consider all finite unions of half-open intervals $[a, b)$ of the extended real line. It forms a field, **F**, and its Borel extension, *BF*, consists of all the Borel sets of the real line. Many of these sets, including those corresponding to single points, are merely ideal events to which nothing corresponds in the outside world.

This philosophical attitude toward the infinite is evident both in the *Grundbegriffe* and in Kolmogorov's later writings. In 1948 he advocates defining probability on measure algebras.[16] The set of elementary events, *E*, disappears, and probability is defined directly on an algebra of "composite events." He writes

> The notion of an elementary event is an artificial superstructure imposed on the concrete notion of an event. In reality, events are not composed of elementary events, but elementary events originate in the dismemberment of composite events.

His 1963 theory of objectively random sequences,[17] which we will discuss in chapter 8, focuses on finite sequences. The modern theory of measure and integral is viewed pragmatically, not as metaphysics but rather as an idealization that allows the infinite to approximate the finite.

SUMMING UP

Kolmogorov's overarching achievement in *Foundations of the Theory of Probability* was to make probability a part of modern mathematics of the infinite. He used the new developments in the theory of measure and integral to construct an abstract framework that put in everything that was needed, left out what was not, and made precise sense of probabilistic concepts that had been informal and somewhat mysterious.

He made sense of random quantities. He gave a general definition of conditional probability that dispelled apparent paradoxes that had arisen from informal argument.

Borel had used infinite sequences and countably infinite additivity to generalize Bernoulli's law of large numbers to the strong law of large numbers. Kolmogorov's framework incorporates countable additivity and allows spaces whose points are whatever you please. Within it he proved an even stronger law of large numbers. He proved an extension theorem that showed how to build up an infinite-dimensional stochastic process from a consistent series of finite dimensional stochastic processes.

With such an abstract and idealized framework, questions must arise as to the connections to reality. It is clear from later work that Kolmogorov worried about these problems. Philosophically he was a finitist. Infinite idealizations were to be constructed in such a way as to be approximable by the finite.

APPENDIX 1. MEASURE OF A COMPLEX SET

What is the area measure of some set of points in the Euclidian plane? Or of length on the line, or of volume in space? Questions about the measure of complicated sets of points had come under consideration in the wake of Cantor's theory of infinite sets.

Peano and Jordan followed the basic strategy that the ancient Bryson of Heraclea[18] used in approaching the problem of squaring the circle. Bryson reasoned that the area of the circle must be greater than any inscribed regular polygon and less than any circumscribed regular polygon. As the number of sides of the polygons increased, the areas became closer and closer. He believed, without stating it explicitly or offering any proof, that in the limit they would coincide. Archimedes approximated the ratio of the circumference of a circle to its diameter, π, by comparing inscribed and circumscribed 96-sided regular polygons.

Peano and Jordan generalized this idea to the concepts of inner and outer content. For example, on the real line, intervals are assigned their length as measure. (This includes points that, as degenerate intervals, are assigned measure 0.) These measures are fundamental, and the concept of measure is extended to other point sets as follows. Consider finite sets of intervals that cover the set of points in question; it is contained in their union. Associate with each such covering set the sum of the lengths of intervals in it. The greatest lower bound of these numbers is called the *outer content* of the set. Working from the other side, consider finite sets of nonoverlapping intervals whose union is contained in the set in question. Associate with each such set the sum of the lengths of its members. The least upper bound of these numbers is called the *inner content* of the set. If the outer and inner content of a point set are equal, then the set is measurable in the sense of Peano and Jordan, and that number is its measure. If not, then the set is *not measurable*—the concept of measure simply does not apply. For instance, the set of rational points in the interval [0, 1] is not Peano-Jordan measurable. Its outer content is 1, while its inner content is 0.

Borel took a stronger approach that made more sets measurable. Borel constructs the Borel measurable sets out of the intervals by countably infinite set-theoretic operations and defines their measure by postulating countable additivity.

A countable union of mutually disjoint intervals has as its measure the infinite sum of the lengths of the intervals. Notice that the set of rationals now is measurable; it has measure zero. That leaves a question hanging. Are all such sets now measurable?

APPENDIX 2. NONMEASURABLE SETS

We present Vitali's 1905 construction of a nonmeasurable set[19] in a probabilistic setting. Consider the unit wheel of fortune, with the points indexed by the real numbers in the half-open interval [0, 1). We assume that this wheel is fair, so that if a measurable set of points is displaced a fixed distance around the circle, we get a set with the same probability.[20] The relation $|x - y|$ *is rational* is an equivalence relation and thus partitions [0, 1) into equivalence classes. For example, every point that we can get to by moving a rational distance from $\frac{1}{4}$ is in the same equivalence class as $\frac{1}{4}$. (This class includes all rational points.) Every point that we can get to by moving a rational distance from $\pi/4$ is in the same equivalence class as $\pi/4$.

Choose one member from each of these equivalence classes to form a choice set* C. For each rational r in [0, 1) let C_r be the set gotten by translating C the distance r around the circle. Since there are a

---

*You could not carry out this infinite process, but presumably God could. The existence of the choice set is guaranteed by the axiom of choice in set theory.

countably infinite number of rationals, these sets form a countably infinite partition of the circle. If they have a probability, they must have the same probability by translation invariance. If it is 0, then the probability of the whole circle adds to 0; if positive, then the probability of the whole circle is infinite (by countable additivity.) So these Vitali sets are nonmeasurable.[21] They cannot have a probability.

Subsequently, other nonmeasurability results were proved. An example with the same flavor was constructed in 3-dimensional Euclidian space by Hausdorff (1914)[22]—and later generalized by Banach and Tarski (1924)[23]—using only finite additivity. The 1-dimensional property of translation invariance is generalized to congruence.

Consider again picking a point at random from [0, 1), and suppose that probability is countably additive. Banach and Kuratowski, in 1929,[24] showed that if Cantor's continuum hypothesis holds, there must be nonmeasurable sets. No invariance principle is assumed.

Thomas Bayes

CHAPTER 6

———

INVERSE INFERENCE

From Bayes and Laplace to Modern Statistics

You are screening new drugs for a certain disease. Some patients get better by at least a certain amount; some don't. For one new drug, more get better on the drug than on a placebo. How confident should we be of the new drug's effectiveness on the evidence? To be explicit, before the experiment, you take the probability that the drug will be effective to be low. It is just one of many drugs being screened for this disease. Given the evidence of your trial, you want to know the probability that the drug is effective—that it raises the patient's chance of improvement by a substantial amount. Is it still low? Is it higher than it was? It is high enough to take seriously? These are questions about the inference from real, often small, data sets to the chances that govern the generation of that data. The fundamental idea that allows us to properly address these questions was first introduced by Thomas Bayes.

Bayes states his own great idea thus:

Given the number of times in which an unknown event has happened and failed:

    Required the chance that the probability of its happening in a single trial lies somewhere between any two degrees of probability that can be named.[1]

This is how he begins his 1763 *Essay towards Solving a Problem in the Doctrine of Chances*. His predecessors, up to and including Bernoulli and de Moivre,* had reasoned from chances to frequencies. Bayes gave a mathematical foundation for statistical inference—inference from frequencies to chances.

Bayes' essay did not appear during his lifetime. It was published two years after his death by his friend Richard Price. Price communicated the essay, together with an introduction and an appendix by himself, to the Royal Society, and it was promptly published in its *Philosophical Transactions* in 1763. Price, referring to an introduction that Bayes himself had written to the *Essay* (and that has not survived) tells us that (Thomas Bayes, "An Essay towards Solving a Problem in the Doctrine of Chances," *Philosophical Transactions of the Royal Society* 53 (1763): 370–418)

> . . . he says that his design at first thinking of the subject of it was, to find out a method by which we might judge concerning the probability that an event has to happen, in given circumstances, upon the supposition that we know nothing concerning it but that, under the same circumstances, it has happened a certain number of times and failed a certain other number of times.

This is a program to find the predictive probability, the probability that something will happen next time, from the past statistics. Price goes on to say that Bayes perceived that this would not be difficult if he first solved the problem that begins the essay. Indeed this is so. We will see that this step was later taken by Laplace in establishing his celebrated rule of succession.

---

*Both Bernoulli and de Moivre claimed to have solved the problem of inverse inference, at least for large numbers of trials, but in fact they did something different. See chapter 4.

Bayes appears to have been motivated not by practical concerns of law or medicine, but rather by questions of mathematical philosophy.

BAYES VERSUS HUME

Price emphasizes the philosophical magnitude of the step—its importance for inductive reasoning:

> Every judicious person will be sensible that the problem now mentioned is by no means a curious speculation in the doctrine of chances, but necessary to be solved in order to a sure foundation for all our reasonings concerning past facts, and what is likely to be hereafter.

Put this way, the project can be seen as an answer to Hume. In the *Enquiry Concerning Human Understanding* (1748), Hume writes

> Though there be no such thing as Chance in the world; our ignorance of the real cause of any event has the same influence on the understanding, and begets a like species of belief or opinion.
>
> . . . Being determined by custom to transfer the past to the future, in all our inferences; where the past has been entirely regular and uniform, we expect the event with the greatest assurance, and leave no room for any contrary supposition. But where different effects have been found to follow from causes, which are to appearance exactly similar, all these various effects must occur to the mind in transferring the past to the future, and enter into our consideration, when we determine the probability of the event. Though we give the preference to that which has been found most usual, and believe that this effect will exist, we must not overlook the other effects, but must assign to each of them a particular weight and authority, in proportion as we have found it to be more or less frequent.
>
> . . . Let any one try to account for this operation of the mind upon any of the received systems of philosophy, and he will be

sensible of the difficulty. For my part, I shall think it sufficient, if the present hints excite the curiosity of philosophers, and make them sensible how defective all common theories are in treating of such curious and such sublime subjects. (Section VI)

There is a very good case made in S. L. Zabell ("Laplace's Rule of Succession," *Erkenntnis* 31 (1989): 283–321.) that Bayes achieved his main results shortly after Hume issued this challenge.

The first bit of evidence is that in 1749, David Hartley published a book, *Observations on Man*, in which he says that "An ingenious friend has communicated to me a solution to the inverse problem . . ." and goes on to describe the problem essentially as stated by Bayes in the beginning of his essay. The second is that a notebook of Bayes' was discovered by Dale (A. I. Dale, "A Newly-Discovered Result of Thomas Bayes'," *Archive for History of Exact Sciences* 35 (1986): 101–13); it contains a result that appears in the essay, sandwiched between entries dated 1746 and 1749.

It is natural for Bayes to be seen as an answer to Hume. There are Humean echoes in Price's appendix, in which Price discusses Hume's example of the rising of the sun. In the *Enquiry* (1749, Section IV), Hume uses it to illustrate skeptical doubts:

> . . . that the sun will not rise tomorrow is no less intelligible a proposition, and implies no more contradiction than the affirmation, that it will rise. We should in vain, therefore, attempt to demonstrate its falsehood.

And in the *Treatise* (Section XI, "Of the Probability of Chances," 1739), he juxtaposes its common sense certainty:

> One would appear ridiculous, who would say, that it is only probable that the sun will rise to-morrow, or that all men must dye; though it is plain that we have no further assurance of these facts, than what experience affords us.

In Price's appendix, as an example of Bayes' result, we find the following ("An Essay towards Solving a Problem in the Doctrine of Chances," *Philosophical Transactions of the Royal Society of London* 53 (1763): 409):

Let us imagine to ourselves the case of a person just brought forth to life into this world and left to collect from his observation of the order and course of events what powers and causes take place in it. The Sun would, probably, be the first object that would engage his attention; but after losing it the first night he would be entirely ignorant whether he should ever see it again.

Price goes on to show that after a million observations, the chance of the sun's rising lies in a small interval close to 1 with high probability. It is clear from this (and from the preceding passage that explicitly mentions uniformity of nature) that Price takes Bayes' essay to be the answer to Hume. The matter is clinched by the title page given by Price to reprints of Bayes' essay (as recently discovered by Stephen Stigler[2]). It is "A Method of Calculating the Exact Probability of all Conclusions Founded on Induction."

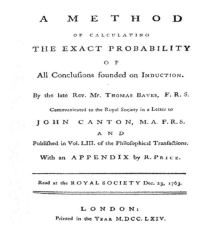

Opening page to Bayes' *Essay*

In fact, Price was a good friend of both Hume and Bayes. He used results of Bayes in his *Four Dissertations*, which he sent to Hume. In the last dissertation he argued that Hume's famous essay on miracles did not give proper weight to the testimony of many witnesses.

Hume replied (R. Klibansky and E. C. Mossner (eds), *New Letters of David Hume* (Oxford: Clarendon Press, 1954): 233–34):

I own to you, that in the Light in which you have put this Controversy, is new and plausible and ingenious, and perhaps solid. But I must have some more time to weigh it, before I can pronounce this Judgment with Satisfaction to myself.[3]

But Hume, although a great philosopher, was not a good mathematician, and it is not likely that he understood Bayes' contribution.

BAYES ON PROBABILITY

The essay begins with a development of probability, with some remarkable anticipations of the modern coherence views that we met in chapter 2. Bayes begins by taking expected value as basic ("An Essay towards Solving a Problem in the Doctrine of Chances," *Philosophical Transactions of the Royal Society of London* 53 (1763): 376):

> The *probability of any event* is the ratio between the value at which an expectation depending on the happening of the event ought to be computed, and the value of the thing expected upon its happening.

The phrase "ought to" may seem strange in this context, but it refers to the correct computation of the expected value of a gamble, with which all readers would have been familiar. He is then saying that a gamble or contract that pays N if event e should have an expected value of $N \cdot P(e)$, or that

$$P(e) = \frac{\text{expected value of the wager } N \text{ if } e}{N}.$$

He follows this definition by saying, "by chance I mean the same as probability." Both words are used subsequently in the exposition, but—as we shall see—both are not to be interpreted in the same way.

It is of some interest that Bayes, in his introduction (which Price has seen but we have not), is reported to apologize for this definition. Bayes apparently did not want to become embroiled in philosophical controversies about the nature of probability. Price tells us, "Instead of the proper sense of the word *probability*, he has given us that which all will allow to be its proper measure in every case where the word is used." (Price's Forward in "An Essay towards Solving a Problem in the Doctrine of Chances," *Philosophical Transactions of the Royal Society of London* 53 (1763): 375). We are not told what Bayes regarded as the proper sense of the word.

On the basis of this definition, Bayes argues for the basic properties of probability.

Additivity of probabilities of disjoint events is gotten from additivity of expectations. If the proper value of a wager giving N if e_1 is a, N if e_2 is b, and N if e_3 is c, and these are mutually exclusive, then the value of N if [e1 or e2 or e3] should be $a+b+c$. Otherwise there is a kind of inconsistency. In particular, if e1, e2, e3 are mutually exclusive and exhaustive, then holding all three wagers together gives N for sure, so the probabilities of the three events should add to 1. The negation rule is noted as a special case. (Notice how close this already is to the Dutch book theorems of the twentieth century that we met in chapter 2.)

Bayes then goes on to establish the definition of conditional probability. He gives a separate treatment for the case where the conditioning event precedes the conditioned one and the case where the conditioning event is subsequent to the conditioned one. The latter case is thought to be more problematic, as the conditioning goes backward in time. This leads to the interesting argument given in his proposition 4. Bayes invites us to consider an infinite number of trials determining the occurrence of the conditioning and conditioned events ("An Essay towards Solving a Problem in the Doctrine of Chances," *Philosophical Transactions of the Royal Society of London* 53 (1763): Proposition 4, p. 379):

> If there be two subsequent events to be determined every day, and each day the probability of the 2nd is b/N and the probability of both P/N, and I am to receive N if both events happen on the 1st day on which the 2nd does; I say, according to these considerations, the probability of my obtaining N is P/b. . . .

Bayes remarks that on the first day either the condition happens—or if not he is facing the same wager as before:

> Likewise, if this coincident should not happen I have an expectation of being reinstated in my former circumstances.

That is to say, the probability of a win on the supposition that E_2 (the second) did not happen on the first day is just the original probability of a win. Let us assume unit stakes, so that expectation equals probability, to simplify the exposition.

Then, letting E_1 be the first event and E_2 the second, he argues as follows:

$$P(\text{win}) = P(\text{win on day 1}) + P(\text{win later})$$

$$= P(E_1 \text{ and } E_2) + P(-E_2) \, P(\text{win})$$

$$= P(E_1 \text{ and } E_2) + [1 - P(E_2)] \, P(\text{win}),$$

$$P(\text{win}) - P(\text{win}) + P(E_2) \, P(\text{win}) = P(E_1 \text{ and } E_2),$$

$$P(\text{win}) = P(E_1 \text{ and } E_2)/P(E_2).$$

That this value is to be taken as the probability of E_1 on the supposition E_2 is taken as a corollary, but the exposition of the corollary contains an interesting twist:

> Suppose, after the expectation given me in the foregoing proposition, and before it is all known whether the first event has happened or not, I should find that the second event has happened; from hence I can only infer that the event is determined on which my expectation depended, and have no reason to esteem the value of my expectation either greater or less than before.

Immediately following, he gives a *money-pump* argument:

> For if I have reason to think it less, it would be reasonable to give something to be reinstated in my former circumstances, and this over and over again as I should be informed that the second event had happened, which is evidently absurd.

He completes this by considering the opposite case:

> And the like absurdity plainly follows if you say I ought to set a greater value on my expectation than before, for then it would be reasonable for me to refuse something if offered on the condition that I relinquish it, and be reinstated in my former circumstances . . .

Bayes already had the idea of coherence arguments (chapter 2) for both unconditional and conditional probabilities!

THE INVERSE PROBLEM AND THE BILLIARD TABLE

With conditional probability in hand, Bayes proceeds to the problem with which he begins the *Essay*. Suppose a coin, about whose bias we know nothing at all, has been flipped n times and has been heads m times. If x is the chance that the coin comes up heads on a single toss, Bayes requires

$$P(x \text{ in } [a, b] \mid m \text{ heads in } n \text{ tosses}).$$

This conditional probability is

$$\frac{P(x \text{ in } [a, b] \text{ and } m \text{ heads in } n \text{ tosses})}{P(m \text{ heads in } n \text{ tosses})}.$$

To evaluate this, Bayes must assume something about the prior probability density over the chances.* He assumes a uniform prior density as the correct quantification of knowing nothing concerning it. Anticipating that this might prove controversial, and of course it has, he later offers a different justification in a scholium. On this basis, he applies Newton's calculus to get

$$\frac{\int_{a}^{b} \binom{n}{m} x^m (1-x)^{n-m} \, dx}{\int_{0}^{1} \binom{n}{m} x^m (1-x)^{n-m} \, dx}.$$

---

*For clarity, we adopt modern terminology here, reversing Bayes' use of probability and chance.

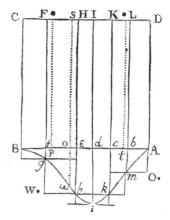

Figure 6.1. Bayes' "Billiard Table" from his *Essay*

How are these integrals to be solved? Bayes evaluates the integral in the denominator by a geometrical trick. This is Bayes' "billiard table" argument.

I throw a red ball at random on a table and mark its distance from the leftmost side. I then toss n black balls one by one on the table, as shown in figure 6.1. Call a ball that falls to the right of the red ball a head and one that falls to the left a tail. This corresponds to choosing a bias at random and flipping a coin of that bias n times. Now nothing hangs on the first ball being the red one. I could just throw $n + 1$ balls on the table and choose the one to be the red ball, the one to set the bias, at random. But if I choose the leftmost ball to be the red one, all black balls count as heads, and if I choose the right one to be the red ball, no black balls count as heads, and so forth. Thus the probability of m heads in n tosses is the same for $m = 0$, $m = 1, \ldots,$ $m = n$. It must be $1/(n + 1)$. This is the value of the integral in the denominator. The integral in the numerator is harder. There is no general closed-form solution. Bayes, however, gives a way of approximating it.

In a scholium, Bayes uses his evaluation of the denominator to argue for his quantification of ignorance. He argues that if I know nothing about the event except that there are n trials, I have no reason to think that it would succeed in some number of trials rather than another. We can take $P(m$ heads in n tosses$) = 1/(n + 1)$ as our quantification of ignorance about outcomes. The uniform prior, in fact, follows from this—although Bayes did not have the proof.

LAPLACE

Bayes' work was not immediately taken up in England.* Instead, the development of inverse inference moved to France, where it was advanced by the brilliant mathematician and astronomer Pierre-Simon Laplace. Laplace's interest in probability was not only theoretical. He was interested in distributions of error in astronomical observation, and the correct method of statistical inference for this application. This may be one of the reasons for the immediate impact of his work. But Laplace also worked on the same problem that Bayes considered—flipping a coin with unknown bias.

Pierre-Simon Laplace

Suppose that heads has happened invariably on every one of n tosses. What is the probability, on Bayes assumption, that it will happen on the next trial? It is the probability of $n + 1$ heads in $n + 1$ tosses

* Indeed, it was ignored in its home country until de Morgan in the 1830s brought the Bayesian approach home.

conditional on n heads in n tosses. By Bayes billiard table argument, this is

$$\frac{1/(n+2)}{1/(n+1)} = \frac{n+1}{n+2}.$$

Using this formula, Laplace, somewhat tongue in cheek, calculates the probability that the sun will rise tomorrow given that it has risen every day for 5000 years. This is a reference to Hume, if not to Bayes.

More generally, using the uniform prior, Laplace calculates the probability of a success on the next trial given m successes in n trials as:

$$\frac{\int_0^1 \binom{n}{m} x^{m+1}(1-x)^{n-m}\, dx}{\int_0^1 \binom{n}{m} x^m (1-x)^{n-m}\, dx} = \frac{m+1}{n+2}.$$

This is Laplace's rule of succession. Notice that for large numbers of trials an application of Laplace's rule is very close to simply taking the relative frequency of heads as one's probability for heads the next time. In this setting, with a lot of data, naive frequentism does not go far wrong. But who, on initially getting two heads, would give probability one on heads the next time?

GENERALIZED LAPLACE

Suppose a flat prior is not appropriate. The coin may be biased, but it is unlikely to be very biased. Perhaps we might want a prior like that shown in figure 6.2. Or it may be more likely to be biased in one direction rather than another, and the appropriate prior would look like that shown in figure 6.3. Can the simplicity and tractability of the Bayes-Laplace analysis be retained? It can. We choose an appropriate prior density proportional to the likelihood:[4]

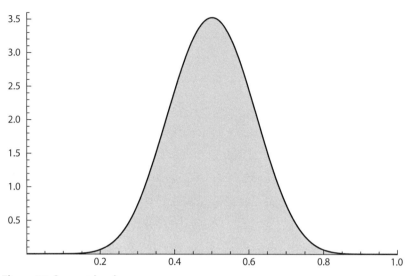

Figure 6.2. Symmetric prior

$$\frac{x^{(\alpha-1)}(1-x)^{(\beta-1)}}{\text{normalizer}},$$

where the normalizer makes the prior density integrate to 1. This is the beta distribution, whose shape is controlled by the parameters alpha and beta. The Bayes-Laplace flat prior has α and β equal to 1. The first example, with the density peaked at about $\frac{1}{2}$, had α and β equal to 10. The second had α equal to 5 and β equal to 10.

Because the prior density is proportional to the likelihood, piling up frequency data keeps the updated density in the beta family. Starting with parameters α and β, a sequence of n trials with m successes and $n-m$ failures, leads to a new beta density with parameters $\alpha+m$, $\beta+(n-m)$. The resulting rule of succession gives the probability of s success for the next trial, on the evidence of m successes in N trials, as

$$\frac{m+\alpha}{n+\alpha+\beta}.$$

It is evident that for large-enough numbers of trials the relative frequency, m/n, again swamps the prior. How fast depends on the magnitudes of α and β.

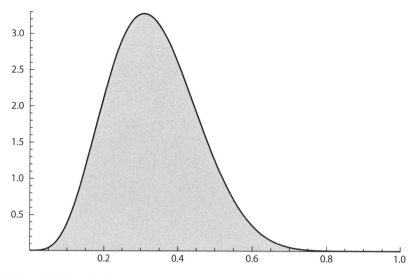

Figure 6.3. Skewed prior

This is true also if we look at not just the predictive probabilities for the next trial, but also at the updated densities. Suppose we have 62 heads in 100 tosses. The updated densities for our uniform, symmetric, and skewed priors are shown in figure 6.4. Disagreement about priors didn't seem to make much difference. Bernoulli's inference from frequency to chance doesn't look too bad here, but now we know what assumptions we had to make to get that result.

There are a limited number of shapes that can be made with beta priors. If you know something about coins, you might want a different shape to quantify your state of prior ignorance. One of us (Persi) knows that coins spun on edge tend to be biased one way or another but more often towards tails. An unknown coin is to be spun. Persi has a bimodal prior density with a somewhat higher peak on the tails' side, which can't be represented by a beta prior. It can, however, be represented by a mixture of two beta densities, one peaked toward heads and one peaked toward tails, with a higher weight on the second. Updating on frequency evidence is still relatively simple, treating the two betas as metahypotheses and their weights as prior probabilities.[5]

More generally, one has a very rich palette of shapes available for quantifying prior states of belief using finite mixtures of betas.

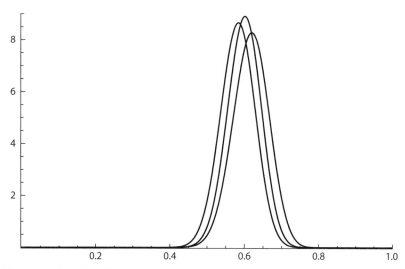

Figure 6.4. Posteriors after 100 tosses

Arguably, you can get anything you might reasonably want to represent your prior mixture of knowledge and ignorance. As before, with a lot of evidence such niceties will not matter much. But if you were going to risk a lot on the next few trials, it would be prudent for you to devote some thought to putting whatever you know into your prior.

CONSISTENCY

Suppose you adopt the point of view that chances are physical properties in the world—there is a true bias of the coin—and that you start with prior beliefs about the chances and feed in frequency evidence to update your degrees of belief. Will you always (or with high probability) learn the true chances? The answer can be no if you start with a pigheaded prior. Suppose you put prior probability one on the coin being biased toward heads. (The integral over the possible biases towards tails is zero.) Then you can never learn that the coin is really biased toward tails.

Let us say that your prior is *consistent* if no matter what the true single-case chance is, you will learn it with chance 1. That is to say that there is chance one of generating an outcome sequence such that it causes your degrees of belief to converge (in the appropriate sense) to the true single case chance. The pigheaded prior of the last paragraph is not consistent, but Bayes' flat prior is. So are all the beta priors and finite mixtures of beta priors that we have discussed. It doesn't take much for a prior to be consistent. Draw any density you want that gives each open interval of chances some positive probability and you have a consistent prior. Nothing hangs on a flat prior. Similar results hold for rolling dice, repeated measurements with normally distributed errors, or indeed for any finite dimensional parametric model.[6] Nonpigheaded priors are consistent. Nothing hangs on a unique quantification of ignorance.

WHY MOST PUBLISHED RESEARCH FINDINGS ARE FALSE

A cookbook frequentist hypothesis tester doesn't have to think. (See the cookbook frequentist in figure 6.5.) He calculates a *p*-value. That is defined as the probability that random noise would generate a false positive.

But a recent attempt to replicate experimental results, published in leading psychology journals, with *p*-values less than 0.05, found that less than half of them replicated.[7] The attempts were done very carefully, with the help of the original authors. There have been similar failures to replicate in other fields. Something like this was predicted by John Ioannidis in a paper entitled, "Why Most Published Research Findings are False."[8] Ioannidis was interested in clinical trials, where false findings may be somewhat more serious than in psychology. Scientists at Amgen trying to replicate "landmark" preclinical results in cancer research found that only 11% replicated.[9] Part of the problem is the mechanical use of *p*-values. As Ioannidis put it,

> Several methodologists have pointed out that the high rate of nonreplication (lack of confirmation) of research discoveries is a consequence of the convenient, yet ill-founded strategy of claiming conclusive research findings solely on the basis of a single study assessed by formal statistical significance, typically for

Figure 6.5. "Did the Sun Just Explode?" XKCD: A Webcomic of Romance, Sarcasm, Math, and Language.

a *p*-value less than 0.05. Research is not most appropriately represented and summarized by *p*-values, but, unfortunately, there is a widespread notion that medical research articles should be interpreted based only on *p*-values.[10]

A Bayesian, looking for the probability that the effect is real given the evidence, would say that the *p*-value is only part of the story. There is the *power* of the test, the probability that a real effect shows up as a positive outcome. There is the prior probability of a real effect, which may depend on the field.

Consider simple examples. Let *e* be the evidence, *T* be a true effect, and –*T* be no true effect. Then on this evidence, the ratio of probabilities of true effect to no true effect is

$$\frac{P(T|e)}{P(-T|e)} = \frac{P(T)}{P(-T)} \cdot \frac{P(e|T)}{P(e|-T)}.$$

Looking only at *p*-values sweeps important considerations under the rug.

In a special issue of *Psychological Science* devoted to replication, Pashler and Harris[11] argue that not unreasonable assumptions about power and priors in psychology combined with a *p*-value of 5% lead to less than even chances of a true effect. In epidemology, where many factors may be screened for relevance to an effect, the situation may be much worse because of the low prior probabilities.

In addition, a desired *p*-value can be actively pursued. Perhaps some runs could be considered as failures; the experiment wasn't working right. Or other experimenters tried, didn't get an effect, and so didn't publish. Sooner or later pure noise will give a good significance level, and that experiment gets published. Or, the experimenter changes the hypothesis to get a good *p*-value from the data. This has come to be called *p*-hacking.[12]

There is now a movement to go beyond mechanical use of *p*-values, convenient as they may be. The proper guidance is Bayesian. Bayes theorem, applied to the totality of evidence, gives the correct method for computing final probability. This has practical consequences. Negative studies should be reported. Multiple studies should be pooled. Prior probabilities (base rates in the field) should not be neglected. The whole likelihood ratio should be reported, not just the *p*-value.

BAYES, BERNOULLI, AND FREQUENCY

From Bayes' point of view, the conclusion of Bernoulli's swindle doesn't look too bad. Given a reasonable prior and lots of data from independent identically distributed trials, it is not unreasonable to infer that the chance is close to the frequency. But a reasonable conclusion doesn't make an argument valid. Bayes showed what was necessary to make the inverse inference from frequency to chance valid.

BAYES CHANGED THE WORLD

Bayes had a philosophical idea that changed the world. We have a prior probability over the chances. We get data and update using Bayes theorem to get a posterior. We put everything we know before getting the data into the prior over what we don't know. Then we feed in data and update. This general scheme applies to more than coin flips or tosses of a die. It applies quite generally. What we don't know may be a chance over vectors, curves, or graphs. Where analytics are intractable, asymptotics or Monte Carlo simulations can be used.

We have already touched on some modern practical applications. Here is another, whose full story is only now emerging. During the Second World War, British analysts broke the German military codes. That this was so remained unknown to the German military, and this was of decisive importance. The German naval enigma codes were deciphered by a group led by Alan Turing, whom we will meet again in chapter 8. These codes had resisted previous code-breaking efforts that relied on frequencies of letters in the code. Turing applied, and in some cases invented, Bayesian techniques—to spectacularly good effect. There was information—where the message came from, time of day, whether it was the length of standard "noise" messages sent to confuse the British, whether messages from the same operator always signed off with a standard-length ending—that played no role in the old deciphering techniques. Turing was able to combine all this information with frequencies in the messages using Bayes' theorem. Some of Turing's work has only recently been declassified.[13] The evolution of western civilization might have taken a different course were it not for this work. Bayes' idea really did change the world.

SUMMING UP

Thomas Bayes—motivated to answer the inductive skepticism of David Hume—solved the fundamental problem of inference from frequency to chance. He filled the gap left by Bernoulli, replacing appeal to moral certainty with probability. The key insight was that one has to put probabilities on the chances and update them by conditioning. This gives us judgment about chances, informed by the evidence of frequencies.

Bayes, and following him Laplace, began with a tractable special case: a uniform prior probability over the chances. Other cases are now tractable in the same way. Different prior beliefs can approach each other when updated with the same evidence of a substantial number of trials. In some cases, with a large number of trials, Bernoulli's swindle gives approximately correct results (but in other cases it doesn't).

We said in the last chapter that Kolmogorov made probability part of mathematics. Bayes made statistical inference part of probability.

Failure to pay attention to all the pieces of a thorough Bayesian analysis may be a mistake.

APPENDIX. BAYESIAN THOUGHTS ON PROBABILITY VERSUS STATISTICS

Bayesians from Bayes to de Finetti and your present authors view probability and statistics as part of the same subject. De Finetti's book was entitled *Probability, Induction and Statistics*![14] Not everyone agrees. A large group of statisticians think that statistics is a separate subject, needing its own foundations.

"Probability is part of mathematics." In a typical probability problem, we are given a probability assignment, $P(x)$, for x in some set, and a subset of outcomes, A, is specified. We are required to compute, or approximate, $P(A)$, the sum of $P(x)$ for all x in A. Since x can vary in a large set and A can be complicated (think of the birthday problem or look in a probability textbook), these include challenging problems that have engaged the best mathematicians for more than 400 years.

"Statistics is the opposite of probability." In a typical statistics problem, we are given a *family* $\mathbf{P} = \{P_1, P_2, \ldots\}$ of probability distributions, shown one x said to be drawn from one of these, and asked to guess at, or estimate, from which P_i x was drawn.

Bayes' great idea was to make statistics a part of probability by putting a prior distribution, π_i, on the various probability distributions, P_i. Then, with x observed, Bayes' theorem says that the posterior probability for the various i's is proportional to

$$P_i(x)\pi_i.$$

One can then choose the i with the largest posterior probability (or the i minimizing some average loss).

The tension is this: Bayesian has to specify π_i, and this can be difficult to do—even in the birthday problem. Is it really the case that birthdays are uniformly distributed over the possible outcomes? What about weekend effects? (20% fewer children are born on days of the weekend than on weekdays.) What about seasonal effects?[15] If you know about these, that knowledge should be reflected in your prior.

There is an enormous statistical literature attempting to do statistics without specifying a prior. The "Einstein" of statistics, R. A. Fisher, suggested maximum likelihood: choose the i that makes the observed data most likely [the maximum of $P_i(x)$]. This is the same as Bayes' rule when π_i is uniform (independent of i).[16] There is a lot of current research on "objective Bayes" trying to generalize this correspondence to more realistic problems with infinite spaces.

A host of other non-Bayesian approaches to choosing i have emerged too: minimum variance, unbiased, chi-square, admissible, minimax, maximin. . . . For each of these, there is a corresponding Bayesian analysis—what prior distribution would one have to assume to have the non-Bayesian estimate be sensible? Sometimes this leads to tension; there are no unbiased Bayes estimates. These discussions led to a fascinating theorem (which horrified its discoverers, Abraham Wald and Charles Stein, who were virulent anti-Bayesians). It says, roughly put, any sensible estimator[17] is Bayes for some prior. This led the biostatistician Jerome Cornfield to conclude

> The Bayesian outlook can be summarized in a single sentence: Any inferential or decision process that does not follow from some likelihood function and some set of priors has objectively verifiable deficiencies.[18] Even more roughly put, *any statistical procedure that is not foolish is Bayesian.*[19]

That is not to say that Bayesian analysis is always easy. Consider flicking an actual tack, counting 1 if it lands point up and 0 if it lands point to the floor. In a classical Bayesian analysis, θ is the chance of point up on a single flick, and this is unknown. Suppose that a uniform prior is used for θ. Suppose further that the tack has been flicked 10 times and has never come point up. What is the chance that it doesn't come point up in the next 10 flicks? The classical calculations show that this is about $\frac{1}{2}\left(\frac{11}{21}\right)$. Suppose 10 is replaced with n; after no success in

n trials, what is the chance of no success in the next n trials? The answer is again about $\frac{1}{2}$: $(n+1)/(2n+1)$ for arbitrarily long initial strings of "no success." If this is an unwelcome surprise, then perhaps the uniform prior is suspect.

Harold Jeffreys and Dorothy Wrinch[20] worried about this,[21] trying on a variety of priors. They suggested putting some prior probability mass at 0 and some at 1. With a prior putting probability $\frac{1}{3}$ at 0, $\frac{1}{3}$ at 1, and uniform in between, the chance of 10 failures given 10 previous failures is more than 90%.[22] This is an example of what I. J. Good (who worked with Turing breaking the Enigma code) called the device of imaginary results.[23] Even if the tack is never flipped, the thought experiment embodied in the calculation shows that the uniform prior is perhaps not appropriate! There is also a useful literature on Bayesian robustness, studying how changes in the prior affect the conclusions.[24]

As said before, the basic Bayesian framework is simple, clean, and consistent. The process of doing real applied statistics is much messier. A readable rich account can be found in David Cox's *Principles of Statistical Inference*.[25] The preliminaries to finding a good model, often called exploratory data analysis, have their own foundations.[26]

Contemplating messy practice, we are reminded of something our good friend Amos Tversky used to say: "You can lie with statistics, but it's a lot easier to lie without them."

Bruno de Finetti

CHAPTER 7

UNIFICATION

Cardano knew how to make inferences from chance to frequencies. Bayes and Laplace then taught us how to make inferences from observed frequencies to chance—all within the framework of probabilistic degrees-of-belief. But what, after all, *is* chance?

We know how to deal with it computationally, but we don't seem to know what it is.

Bruno de Finetti tells us not to worry: *Chance does not exist.*

Yet, it is perfectly legitimate to carry on just as if it did! How is this possible? This is de Finetti's great idea, and the story of this chapter.

That probability as a physical quantity—physical propensity, objective chance—does not exist is a position that has been taken by many philosophers, most notably David Hume. But de Finetti did more than just take this philosophical stance. He showed that, in a certain

precise sense, that if we dispense with objective chance *nothing is lost*.
The mathematics of inductive reasoning remains exactly the same.

Consider flipping a coin of fixed bias. The trials are assumed to be
independent, and as a result, they exhibit another important property. Order doesn't matter. To say that order doesn't matter is to
say that if you take any finite sequence of heads and tails and per-
mute the outcomes any way you please, and the resulting sequence has
the same probability. We say that this probability is invariant under
permutations.

Or, to put it another way, the only thing that does matter is relative frequency. Outcome sequences that have the same frequencies of
heads and tails have the same probability. Frequency is said to be a sufficient statistic. To say that order doesn't matter or to say that the only
thing that does matter is frequency are two ways of saying exactly the
same thing. This property is called *exchangeability* by de Finetti.

Now suppose a coin with unknown bias is flipped 3 times with a
uniform prior over bias, as in Bayes and Laplace. An outcome sequence
with two heads and a tail can be gotten three ways:

$$HHT,$$

$$HTH,$$

$$THH.$$

Initial probabilities on the first trial are $\frac{1}{2}$, and probabilities on subsequent trials are gotten by using Laplace's rule of succession (chapter 6):

$$P(\text{HHT}) = \left(\tfrac{1}{2}\right)\left(\tfrac{2}{3}\right)\left(\tfrac{1}{4}\right),$$

$$P(\text{HTH}) = \left(\tfrac{1}{2}\right)\left(\tfrac{1}{3}\right)\left(\tfrac{2}{4}\right),$$

$$P(\text{THH}) = \left(\tfrac{1}{2}\right)\left(\tfrac{1}{3}\right)\left(\tfrac{2}{4}\right).$$

These are the same. The denominators, 2, 3, and 4, have to be the
same because of the rule of succession. The numerators have to be
the same (though not in the same order) because of the same frequencies of heads and tails. The point is that here we have an example of
exchangeability without independence. In fact, the uniform prior was
not essential. Any uncertainty about the bias of the coin in independent trials gives exchangeable degrees of belief.

De Finetti proved the converse. Suppose your degrees of belief—the judgmental probabilities of chapter 2—about outcome sequences are exchangeable. Call an infinite sequence of trials exchangeable if all of its finite initial segments are. De Finetti proved that every such exchangeable sequence can be gotten in just this way. It is just *as if* you had independence in the chances and uncertainty about the bias. It is just *as if* you were Thomas Bayes.

What the prior over the bias would be in Bayes is determined in the representation. Call this the imputed prior probability over chances the *de Finetti prior*. If your degrees of belief about outcome sequences have a particular symmetry, *exchangeability*, they behave *just as if* they are gotten from a chance model of coin flipping with an unknown bias and with the de Finetti prior over the bias.

So it is perfectly legitimate to use Bayes' mathematics even if we believe that chance does not exist, as long as our degrees of belief are exchangeable.

De Finetti's theorem helps dispel the mystery of where the prior belief over the chances comes from. From exchangeable degrees of belief, de Finetti recovers both the chance statistical model of coin flipping and the Bayesian prior probability over the chances. The mathematics of inductive inference is just the same. If you were worried about where Bayes' priors came from, if you were worried about whether chances exist, you can forget your worries. *De Finetti has replaced them with a symmetry condition on degrees of belief.* This is, we think you will agree, a philosophically sensational result.

READING DE FINETTI

Suppose you are interested enough to want to go to the source. We recommend chapter 9 of de Finetti's *Probability, Induction, and Statistics*[1] as the best stand-alone account of his ideas and his program. This includes a historical and comparative survey, which moves from the beginnings to the crisis of the classical formulation, the rise of objectivistic concepts, the erosion of the objectivistic positions, critical examination of controversial aspects, and reconstruction of the classical formulation according to the subjectivistic viewpoint. Although de Finetti is never easy to read, this is the best we know.

A longer, readable essay is his summing up of earlier work in his 1937 "La Prévision," which is available in a translation that he himself oversaw: "Foresight: Its Logical Laws, Its Subjective Sources."[2] Just a year later he published our favorite paper, "Sur la condition d'équivalence partielle," giving wide generalizations of his earlier work in the context of a deep philosophical discussion of inductive reasoning.[3, 4] We will discuss these generalizations later in this chapter.

Returning to de Finetti's original theorem, it is possible to approach it in a more simple and transparent way if we start with finite sequences. This is next up.

FINITE EXCHANGEABLE SEQUENCES

One thing one might worry about—de Finetti certainly did—is the reliance on infinity. De Finetti's theorem is a limiting result; it is not true for exchangeable probabilities on finite sequences.[5] But it can be approximated on finite sequences. This fact can be used to prove de Finetti's theorem for infinite sequences.

Suppose that a coin is flipped twice. Outcomes can be HH, HT, TH, or TT. The set of possible probabilities over these outcome sequences consists of the points in a tetrahedron, with vertices giving each outcome probability 1. Exchangeability requires that HT and TH have the same probability and thus determines a plane slicing through this tetrahedron, as shown in figure 7.1. The probability $(0, \frac{1}{2}, \frac{1}{2}, 0)$—indicated on the upper right in the figure—that gives HT and TH each the probability $\frac{1}{2}$ is not independent. If you see a head on the first toss, you will be sure of a tail on the second. This is just the probability that you would get for sampling from an urn with one red and one black ball without replacement.

The exchangeable probabilities that make the tosses independent are gotten from the surface of independence (called the Wright manifold in population genetics) within the plane of exchangeability. The points on the plane and beneath the curve in figure 7.2 are exchangeable probabilities that can be represented as averages over independent

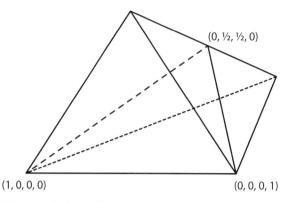

Figure 7.1. Exchangeable sequences

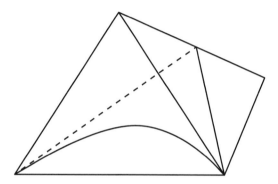

Figure 7.2. Averages of independent probabilities

probabilities. Those above the curve on the plane are exchangeable probabilities that cannot be so represented.

The exchangeable sequences above the curve can, however, be represented as averages of the kinds of probabilities we get from sampling from an urn of known composition, with two balls and without replacement. These are the vertices of the exchangeable triangle: 2 red balls (2 heads), 2 black balls (2 tails), and 1 of each. The vertices are extreme points. They cannot be represented as nontrivial averages of other points. Every other point in the triangle can be represented as an average of them.

This generalizes to longer sequences and higher-dimensional geometric objects. You can take it on faith and skip this paragraph or persist, as you please. Consider the vertices of the analogue to the triangle when there are more trials. (This is called the exchangeable simplex.) The vertices of this simplex are probabilities that represent sampling from an urn of the appropriate size without replacement. Suppose that there are M red balls and L black balls in an urn. Sampling without replacement must give a sequence M red balls and L black balls in some order. All these outcome sequences have the same probability, so the probability is exchangeable. Sampling without replacement gives exchangeability. Call the exchangeable probability just gotten e, and suppose for *reductio* that it is a nontrivial mixture of two other probabilities, a and b. Both a and b must give probability 0 to outcome sequences with other than M red and L black balls. But a and b must give each outcome sequence with M red and L black balls the same probability because they are exchangeable. So $a = b = e$. It is not a nontrivial average. It is a vertex of the exchangeable simplex.

Now, *what if you think that there might be another trial*—there might be another ball in the urn? And what if you maintain your belief that order doesn't matter under that supposition? That is to say that your beliefs should be extendable to exchangeable beliefs on three trials. Then the point $(0, \frac{1}{2}, \frac{1}{2}, 0)$ at the top vertex of our triangle is no longer an option. [Why? It gives probability 0 to HH and to TT. Then its extension to three trials must give probability 0 to HHH, HHT, TTH, and TTT. If it were exchangeable, it would have to give probability 0 to any sequence with 2 heads or 2 tails in any position. But how am I to preserve probability of HT being $\frac{1}{2}$, when both its extensions, HTH and HTT must have probability zero?]

Extendability to an exchangeable probability on 3 tosses cuts off nonindependent probabilities near the top vertex of the triangle, as shown in figure 7.3.

Extendability to more tosses further constrains the exchangeable probabilities, with them approximating the mixtures of independent ones, just as sampling without replacement from a large urn approximates sampling with replacement. The appendix to this chapter gives a careful statement (with an error term) for this finite version of the theorem.

What does this show? *If your judgment of symmetry—that probabilities don't depend on order of trials—doesn't itself depend on the number of trials, then you essentially have de Finetti's theorem.*

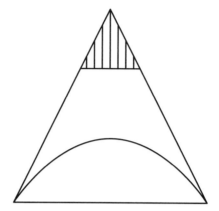

Figure 7.3. Exchangeability extendable to three trials

DE FINETTI'S THEOREM FOR
MORE GENERAL OBSERVABLES

The discussion above has been for binary observables—things like heads and tails, men and women, or 0s and 1s. The whole picture generalizes to essentially arbitrary observables: {red, white, blue}, or real values {successive measurements of length}, or a day-to-day weather maps. Consider successive measurements in an arbitrary space, X. Suppose that we have assigned probabilities to possible outcomes. Thus for A_1, A_2, A_3, subsets of X,

$$P(A_1, A_2, A_3)$$

stands for our probability that the first outcome is in A_1, the second, in A_2, and the third, in A_3. P is *exchangeable* if order doesn't matter. Thus,

$$P(A_1, A_2, A_3) = P(A_2, A_1, A_3) = P(A_2, A_3, A_1)$$
$$= P(A_1, A_3, A_2) = P(A_3, A_1, A_2) = P(A_3, A_2, A_1).$$

This kind of symmetry is supposed to hold for any number of observations. Suppressing measure theoretic niceties, de Finetti's theorem holds:

$$P(A_1, A_2, \ldots, A_n) = \int P(A_1) P(A_2) \cdots P(A_n) \, \mu(dP).$$

Moreover, μ is determined as the limiting proportion of outcomes in a given set.

When X is a 2-point set, for example, $X = \{0, 1\}$, this theorem specializes to the binary version of de Finetti's theorem stated before. In general, the integral is over a terrifyingly large set, the set of all probabilities on X. When $X = \{0, 1\}$, a probability is specified by a number p (the chance of 1), and the integral becomes

$$\int_0^1 p^k (1-p)^{n-k} \, \mu(dp).$$

For 3-valued observables, $\{r, w, b\}$, for example, we get the probability of an outcome sequence:

$$P(r, r, w, r, b, b, w, r) = \int p_r^4 p_w^2 p_b^2 \mu(dp),$$

where the integral is over the set of p_r, p_w, p_b, which are nonnegative and sum to 1.

The general form of de Finetti's theorem looks marvelous until you think about it a bit. It says that we have to think about a probability μ on the set of all probabilities. If X is large, this can be very large indeed. There is a large, esoteric literature about this, but it is not easy to do.[6]

This leads to the question: what do we have to add to exchangeability to get down to mixtures of the day-to-day families of statistics—the normal Poisson, uniform and other standard families? This, too, is a healthy research area. We will give a brief account for the normal next.[7]

DE FINETTI'S THEOREM FOR
THE NORMAL DISTRIBUTION

The normal distribution is the most widely used family of distributions in applied statistical work. It has two parameters, the mean, μ, and the variance, σ^2.

$$f(x \mid \mu, \sigma^2) = \frac{1}{\sigma\sqrt{2\pi}} \, e^{-(x-\mu)^2/2\sigma^2}.$$

Its probability density is the familiar bell-shaped curve shown in figure 7.4. Standard statistics problems begin: let $X_1, X_2, \ldots, X_n$ be independent normal variables with mean μ and variance σ^2. A Bayesian treatment needs a prior density for μ and σ^2. Then

$$P(X_1 \leq x_1, \ldots, X_n \leq x_n) = \iint \Phi_{\mu,\sigma^2}(x_1) \cdots \Phi_{\mu,\sigma^2}(x_n) \, \upsilon(d\mu, d\sigma^2)$$

with $\Phi(x)$ being the area to the left of x under the normal curve.

Can we find a symmetry characteristic of this setup, without ever mentioning the bell-shaped curve? The answer is a bit mathematical. Here is a leading special case. Suppose we know that the mean is 0, so that the only parameter left is the variance, σ^2. For this case, de Finetti's theorem (proved by David Freedman) is as follows.

Theorem: Let $X_1, X_2, X_3, \ldots$ be an infinite sequence of real-valued random variables that are exchangeable *and* satisfy the following, somewhat less than transparent, symmetry:

$$P(X_1 \leq x_1, \ldots, X_n \leq x_n) = P\left(\frac{(X_1 + X_2)}{\sqrt{2}} \leq x_1, \right.$$

$$\left. \frac{(X_1 - X_2)}{\sqrt{2}} \leq x_2, X_3 \leq x_3, \ldots X_n \leq x_n \right).$$

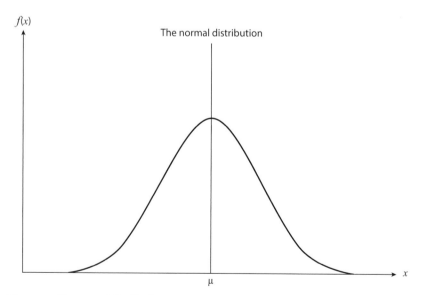

$f(x)$

The normal distribution

μ

x

Figure 7.4. The normal distribution

Then there exists a unique probability v such that

$$P(X_1 \leq x_1, \ldots, X_n \leq x_n) = \int_0^\infty \Phi_{\sigma^2}(X_1) \cdots \Phi_{\sigma^2}(X_n)\, v(d\sigma^2).$$

Similar characterizations are available for most standard families. The challenge of finding more natural symmetry characterizations is ongoing. It seems a lot to ask. The usual justification of the normal distribution is via the central limit theorem. There are Bayesian versions of the central limit theorem, but blending these with de Finetti theory is still for the future.

MARKOV CHAINS

So far, so good, but you might have another worry—which de Finetti also shared. What if your degrees of belief are not exchangeable? *What if order makes a difference?* The simplest sort of patterns in the data stream that we might want to project are those in which the outcome of a trial tends to depend on that of the preceding trial. Exchangeable degrees of belief cannot project such patterns.

Already in 1938, de Finetti suggested that exchangeability needed to be extended to a more general notion of partial exchangeability. The idea was that where full exchangeability fails, we might still have some version of conditional exchangeability. With respect to the kind of pattern just considered, the relevant condition would consist of the outcome of the preceding trial. The notion desired here is that of *Markov exchangeability.*

We are considering a loosening of the assumption of pattern-lessness in the data stream to one where the simplest types of patterns can occur; that is, those where the probability of an outcome can depend on the probability of the previous outcome. One way to say this is that we loosen the assumption that the imputed chances make the trials independent to the assumption that the true chances make the trials *Markov dependent.* Here we can replace the example of the coin flip with that of the Markov thumbtack. A thumbtack is repeatedly flicked as it lies. It can come to rest in either of two positions: point up (PU) or point down (PD). The chance of the next state may well depend on the previous one. Thus there are unknown transition probabilities:

$$PU \quad PD$$

$$PU \quad P(PU|PU) \quad P(PD|PU)$$

$$PD \quad P(PU|PD) \quad P(PD|PD),$$

which an adequate account of inductive inference should allow us to learn.

A stochastic process is *Markov exchangeable* if sequences of the same length having the same transition counts and the same initial state are equiprobable. David Freedman showed that any stationary Markov exchangeable process is representable as a mixture of stationary Markov chains.[8] For another form of de Finetti's theorem for Markov chains, we need the notion of a recurrent stochastic process. A state of a stochastic process is called *recurrent* if the probability that it is visited an infinite number of times is 1. The process is recurrent if all its states are recurrent. Diaconis and Freedman[9] show that a *recurrent* Markov exchangeable stochastic process has a unique representation as a mixture of Markov chains.

MORE PARTIAL EXCHANGEABILITY

De Finetti also had other sorts of cases of partial exchangeability in mind in 1938. He conceptualizes the question of partial exchangeability in general as modeling degree of *analogy* between events:

> But the case of exchangeability can only be considered as a limiting case: the case in which this "analogy" is, in a certain sense, *absolute* for all the events under consideration. . . . To get from the case of exchangeability to other cases which are more general but still tractable, we must take up the case where we still encounter "analogies" among the events under consideration, but without attaining the limiting case of exchangeability.[10]

For his simplest example, he takes the case in which two odd-looking coins are flipped. If the coins look exactly alike, we may take a sequence consisting of tosses of both to be exchangeable. If they look similar, we will want an appropriate form of partial exchangeability, where a toss of coin A may give us some information about coin B, but not as much as we would get from a toss of coin B. Later, he discusses a more interesting, but essentially similar, case in which animal trials of a new drug are partially exchangeable with human trials.

The trials of coin A are exchangeable, as those are of coin B, but they are not exchangeable with each other. One way to say this is to say that for a mixed sequence of trials of both coins, initial segments of the same length that have both the same numbers of heads on coin A and the same number of heads on coin B are equiprobable.

This case, that of Markov exchangeability, and many others can be brought under a general theory of partial exchangeability in which the concept of a *sufficient statistic* plays a key role. Sequences with the same value of a sufficient statistic have the same probability. For simple coin tossing, the statistic is the frequency of heads; for tossing several coins, it is the vector of frequencies of heads of the respective coins; for the Markov case, it is the vector of initial state and transition counts. In each case, under appropriate conditions we have (1) *a de Finetti–type representation* of degrees of belief as mixtures of the extreme points of the convex set of probabilities for which that is a sufficient statistic and (2) *a convergence result* to the effect that learning from experience will with probability one converge to one of these extreme points.

SUMMING UP

Bayes was the father of *parametric* Bayesian analysis. He analyzed a chance model—the sequence of independent and identically distributed coin flips. There is an unknown parameter—the bias of the coin. One puts a degree of belief prior on the parameter, and then one can reason inversely from data to a posterior degree of belief and predictively to anticipated new data. Although he, himself, did not put it in this way, he showed how chance, frequency, and degree of belief all interact to give statistical inference.

De Finetti was the father of *subjective* Bayesian analysis. He took the old idea of symmetry and applied it to degrees of belief. He showed how the elements of Bayes' chance model could all be seen as artifacts of such a symmetry, exchangeability. For cases where we do not have exchangeability, de Finetti showed the same idea applied to weaker symmetries: Markov exchangeability and more general forms of partial exchangeability. Chance functions as a kind of place marker for the appropriate symmetry.

De Finetti showed us how to reason about chance if chance does not exist.

APPENDIX 1. ERGODIC THEORY AS A GENERALIZATION OF DE FINETTI

In an appendix to chapter 4 we briefly visited the 1931 ergodic theorem of George Birkhoff [11] as a general way of establishing a connection between probability and frequency. It is now appropriate to revisit it as a far-reaching generalization of de Finetti's theorem.

What is a symmetry? It is a feature that is invariant under a group of transformations. This is given a lucid explanation in a classical book by Hermann Weyl.[12] Think of invariance under rotations or reflections as marks of familiar physical symmetries. More generally, we have a set of states and a set of measurable subsets of it. A probability measure is *invariant* with respect to a group (or semigroup) of transformations, if every measurable set, A, has the same probability as its inverse image, $T^{-1}(A)$. As an especially important case of symmetry, stochastic processes invariant with respect to time shifts are said to be stationary.

A measurable set is *invariant* with respect to the transformations if the transformations take it into itself (except of a set of probability zero.) The probability measure is *ergodic* with respect to the transformations if the invariant sets either have probability one or probability zero.

The ergodic decomposition theorem says that every invariant probability can be represented as an average (mixture) of ergodic ones. In particular, the ergodic decomposition tells us that stationary processes are representable as mixtures of ergodic processes. An exchangeable probability on a sequence of two-valued random variables—one that is invariant under finite permutations of trials—is stationary. The ergodic probabilities of the decomposition make the coin flips independent and identically distributed. This is the de Finetti representation theorem all over again.

David Freedman's 1962 generalization of de Finetti's theorem to stationary Markov-exchangeable processes also uses ergodic representation.[13]

David Freedman

Freedman, in fact, proves a much more general result. A stationary stochastic process characterized by a kind of sufficient statistic is a mixture of ergodic measures characterized by that statistic.

We see now de Finetti's point of view has a far-reaching generalization to probabilistic symmetries. The ergodic measures can be thought of as surrogates for the physical chance hypotheses. Just as in the coin-flipping case, we reason *just as if* we were reasoning about objective chance—but this is all cashed out as symmetry in our degrees of belief.

APPENDIX 2. DE FINETTI'S THEOREM
ON EXCHANGEABILITY

De Finetti's ideas connecting exchangeability and long-term frequency are such a basic part of the foundations of chance that it
behooves us to work them out in more detail. This appendix gives
a careful statement of de Finetti's theorem in both the 0/1 case
and the most general case. De Finetti's theorem is a limit theorem;
infinite exchangeable sequences are required. However, there are perfectly useful versions for finite exchangeable sequences and useful
quantitative estimates of the accuracy of the infinite limits for finite
sequences. These finite-to-infinite results give proofs for the most
general version of de Finetti's theorem. It all boils down to a simple idea: if you have an urn with many balls, some red and some
white, and you take a sample of a small number, the results will
hardly depend on whether your sample is with or without replacement. Making this precise takes a bit of notation and a bit of math.
However, all that is needed is the classical birthday problem, which
we treated in chapter 1.

DE FINETTI'S THEOREM FOR TWO-VALUED OUTCOMES

Consider binary outcomes. We will call them 0/1, but they might be
H/T for heads and tails or M/W for men and women (as the outcomes of successive births in a family or hospital). We assume, after
reflection to be sure, that probabilities have been coherently assigned
to potential sequences, for example, $P(0), P(1), P(0, 1), P(0, 1, 0), \ldots$.
The assignment $P(\cdot)$ is *exchangeable* if for any $r \geq 1$ and all possible
sequences of r 0s and 1s, $e_1, e_2, \ldots, e_r$, say, P doesn't change if the e_i are
permuted. Thus

$$P(01) = P(10) \quad \text{when } r = 2,$$

$$P(011) = P(101) = P(110), \quad P(001) = P(010) = P(100) \quad \text{when } r = 3,$$

and so on. The order doesn't matter.

It might be that P is only defined for a fixed r (e.g., $r = 100$). Say P is *extendable* if for any $R > r$ there is a $\tilde{P}$ defined on sequences of length R which is exchangeable and restricts to P for its first r coordinates. Thus more data of similar type can be contemplated. De Finetti's theorem now applies:

Theorem (de Finetti's theorem for two values): Let P be an exchangeable and extendable probability assignment to sequences of two values. Then there exists a unique prior probability μ on $[0, 1]$ such that for every n and all sequences $e_1, e_2, \ldots, e_n$,

$$P(e_1, \ldots, e_n) = \int_0^1 \theta^s (1 - \theta)^{n-s} \, \mu(d\theta), \tag{1}$$

with $s = e_1 + \cdots + e_n$, the number of 1s among $\{e_1, e_2, \ldots, e_n\}$. Moreover P assigns probability one to the event

{proportion of ones in the first n places has a limit θ} (2)

and

$$P(\theta \leq x) = \mu(0, x]. \tag{3}$$

Informally, a Bayesian with an exchangeable P is sure that long-term frequencies exist (2) and that P may be represented as a mixture of coin tossing (1). The mixing distribution μ is uniquely identified as the long-term frequency distribution (3). An explanation and proof of de Finetti's theorem is given later in this appendix. It follows as a special case of the general de Finetti theorem, our next topic.

GENERAL FORM OF DE FINETTI'S THEOREM

Observables of interest can be much richer than 0s and 1s. Rolling a die leads to outcomes $\{1, 2, 3, 4, 5, 6\}$. The lengths of successive newborns can be any real number. Yearly temperature profiles give rise to random curves. In general, let $\mathcal{X}$ be *any* set of possible outcomes. A sequence of $\mathcal{X}$-valued outcomes can be assigned probabilities. For example, if A, B, C are subsets of $\mathcal{X}$, $P(A, B, C)$ is interpreted as the probability assigned to the first outcome is in A, the second outcome is in B, and the third outcome is in C. An exchangeable assignment entails assigning the same probability to all arrangements. In the example,

$$P(A,B,C) = P(A,C,B) = P(B,A,C) = P(B,C,A)$$
$$= P(C,A,B) = P(C,B,A).$$

Such assignments are often natural. Think about the lengths of the first three births in a large hospital (starting just past midnight) with

$$A = \{\text{length} \leq 20 \text{ inches}\}, \quad B = \{\text{length} > 18 \text{ inches}\},$$
$$C = \{\text{length between 16 and 25 inches}\}.$$

De Finetti's theorem applies to essentially all such probabilities.*

Theorem (General version of de Finetti's theorem): Let P be an exchangeable, extendable probability on sequences of outcomes in a set $\mathcal{X}$. Then there exists a unique prior distribution μ such that for any n and any sequence of subsets $A_1, A_2, \ldots, A_n$,

$$P(A_1, A_2, \ldots, A_n) = \int_{\mathcal{P}} F(A_1)F(A_2) \cdots F(A_n)\, \mu(dF). \qquad (4)$$

Moreover, for any subset A, P assigns probability one to the outcome

$$\{\text{proportion of first } n \text{ trials in } A \text{ has a limit } \ell\} \qquad (5)$$

with

$$\mu(L) = \text{chance assigned by } P \text{ to the limiting ratio } \ell \text{ being in } L. \qquad (6)$$

Remark: Let us try to explain some of the symbols in (4), (5), and (6). It might help to compare with the parallel equations (1), (2), and (3). To begin with, P on the left side of (4) is the given exchangeable probability. On the right side of (4), $\mathcal{P}$ is the set of all probabilities on $\mathcal{X}$ and μ is a probability distribution on $\mathcal{P}$ (a probability on probabilities!). In the 0/1 case of the first section, the set of probabilities on $\mathcal{X} = \{0, 1\}$ can be identified with $[0, 1]$, with θ in $[0, 1]$ being the chance of 1 on the next trial.

---

*Technically, $\mathcal{X}$ must be a complete, separable metric space but such details will be suppressed.

In the present scenario, F is a fixed probability on $\mathcal{X}$ and (4) says that any exchangable P can be uniquely represented as a mixture of independent and indentically distributed trials with common distribution F. The mixing measure μ is a prior on F. It depends on P, as (5), (6) show. In nonparametric Bayesian statistics, one often defines P by specifying μ, hopefully after a good deal of reflection. This can be a serious undertaking, but it is a hot topic on both the research and applied frontiers.[*][†]

In this chapter, we showed that theorems such as (4), (5), and (6) can fail if only finite exchangeable sequences are considered. It turns out that there is a perfectly useful finite version of de Finetti's theorem; this is developed in the next section. The following sections show how to take limits in the finite theorems to get the infinite theorems.

FINITE VERSIONS OF DE FINETTI'S THEOREM

Begin with the two-valued case. Fix $n \geq 2$ and consider an exchangeable assignment,

$$P(e_1, e_2, \ldots, e_n), \qquad e_i \ 0 \text{ or } 1.$$

The following argument shows that P can be uniquely represented as a mixture of "urn measures." Consider an urn containing n balls with r balls labeled 1 and $n - r$ balls labeled 0. If the urn is well mixed and the balls are removed one by one in random order, a sequence of length n results, containing r 0s and $n - r$ 1s. Call this way of generating sequences P_r: thus

$$P_r(e_1, \ldots, e_n) = \begin{cases} \frac{1}{\binom{n}{r}} & \text{if } e_1 + \cdots + e_n = r \\ 0 & \text{otherwise.} \end{cases}$$

---

[*]J. K. Ghosh and R. V. Ramamoorthi, *Bayesian Nonparametrics* (Berlin: Springer, 2003).

[†]P. Diaconis and D. Freedman (1986). "On the Consistency of Bayes Estimates," *Annals of Statistics* 14 (1986): 1–26.

Theorem (de Finetti's theorem for finite binary sequences): Let P be an exchangeable probability on binary sequences of length n. Then there exists a unique prior distribution μ on $\{0, 1, \ldots, n\}$ such that for any $e_1, \ldots, e_n$,

$$P(e_1, \ldots, e_n) = \sum_{r=0}^{n} P_r(e_1, \ldots, e_n)\, \mu(r). \tag{7}$$

Moreover,

$$\mu(j) = P\{j \text{ ones}\}. \tag{8}$$

Proof: The argument is so simple, we can give it in a line; it's just the law of total probability. For any set of sequences A,

$$P(A) = \sum P(A \mid \text{no. 1s} = r)P\{\text{no. 1s} = r\}.$$

By exchangeability, given r heads out of n, P assigns equal probability to all sequences with r ones. Thus

$$P(A \mid \text{no. 1s} = r) = P_r(A).$$

The proof is completed by simply defining $\mu(j) = P\{\text{no. 1s} = j\}$, giving (7) and (8).   QED

There is nothing special about the balls in the urn being labeled 0/1. Let $\mathcal{X}$ be any set; now we really mean any set, real numbers, vectors, curves, any set. Suppose an urn u contains n balls, each marked by one or another of the set $\mathcal{X}$ (repeats allowed). Let H_u be the probability distribution of n draws made at random *without* replacement from u and M_u be the probability distribution of n draws made at random *with* replacement; H stands for hypergeometric, M for multinomial. These multinomials will enter in the following section.

Theorem (Finite form of de Finetti's theorem, general case): Let $\mathcal{X}$ be a set and P an exchangeable probability on sequences of length n from $\mathcal{X}$. Then there exists a unique probability $\mu(u)$ over urns such that

$$P(A) = \sum_{u} H_u(A)\mu(u), \qquad A \text{ any set of sequences.} \tag{9}$$

In (9),

$\mu(u)$ is the chance that P assigns to sequences resulting in u.   (10)

The proof is precisely as in the binary case, using the law of total probability. Technically, the sum should be replaced by an integral, if $\mathcal{X}$ is infinite. Again we suppress such niceties; see Diaconis and Freedman* for more details.

These theorems give unique representations à la de Finetti: the most general exchangeable probability is a mixture of simple sampling distributions. Because samples with and without replacement are close, this will lead to exchangeable probabilities being close to a mixture of multinomials. We turn to making this precise.

EXPLICIT BOUNDS IN FINITE THEOREMS

Let $\mathcal{X}$ be any set and $\mathcal{P}$ the set of all probabilities on $\mathcal{X}$ (for measure-theoretic details, see Diaconis and Freedman). For $F \in \mathcal{P}$, let F^k be the independent probability on sequences of length k, so $F^k(A_1, \ldots, A_k) = F(A_1)F(A_2) \cdots F(A_k)$ for A_i subsets of $\mathcal{X}$. If μ is a probability on $\mathcal{P}$, let

$$P_{\mu k}(A) = \int_{\mathcal{P}} F^k(A) \, \mu(dF). \tag{11}$$

Let P be an exchangeable probability on sequences of length n. For $1 \leq k \leq n$, let P_k be the marginal distribution on the first k coordinates. Thus

$$P_k(A_1, \ldots, A_n) = P(A_1, \ldots, A_k, \mathcal{X}, \ldots, \mathcal{X}).$$

Theorem: Let $\mathcal{X}$ be a set and P an exchangeable probability on sequences of length n. Then there exists a probability μ on $\mathcal{P}$ such that for all $k \leq n$ and any set A,

$$\left| P_k(A) - P_{\mu k}(A) \right| \leq \frac{k(k-1)}{2n}.$$

*P. Diaconis and D. Freedman, "Finite Exchangeable Sequences," *Annals of Probability* 8 (1980): 45–64.

Thus, if n is large with respect to k^2, the first k coordinates of an exchangeable probability are uniformly well approximated by a mixture of independent identically distributed probabilities. De Finetti's theorem holds approximately! The result is often used in reverse: suppose P is an exchangeable probability on sequences of length k that can be extended to an exchangeable probability on sequences of length n (we can imagine getting more, similar data). Then P is almost a mixture of independent identically distributed probabilities.

The proof is easy and instinctive. Recall the multinomial and hypergeometric probabilities M_u and H_u generated by drawing with or without replacement from an urn u containing n balls labeled with various elements of $\mathcal{X}$ (repeats allowed). For $1 \le k \le n$, let M_{uk}, H_{uk} be the probabilities induced on sequences of length k.

Lemma: For any set A and any urn u,

$$|M_{uk}(A) - H_{uk}(A)| \le \frac{k(k-1)}{2n}.$$

Proof: It is without loss of generality to take $\mathcal{X} = \{1, 2, \ldots, n\}$ and $u = \{1, 2, \ldots, n\}$. Then for any sequence $x = (x_1, \ldots, x_k)$ in $\mathcal{X}$,

$$M_{uk}(x) = \frac{1}{n^k},$$

$$H_{uk}(x) = \begin{cases} \dfrac{1}{n(n-1)\cdots(n-k+1)} & \text{if all } x_i \text{ distinct} \\ 0 & \text{otherwise.} \end{cases}$$

It follows that the worst case for A is $A = \{x: \text{ all } x_i \text{ distinct}\}$. Since $H_{uk}(A) = 1$ and $M_{uk}(A) = n(n-1) \ldots (n-k+1)/n^k$. For this A, direct computation shows

$$|M_{uk}(A) - H_{uk}(A)| = 1 - \frac{n(n-1)\cdots(n-k+1)}{n^k}.$$

Using $(1-x)(1-y) = 1-x-y+xy \geq 1-x-y$ (if $x, y > 0$), we see

$$\frac{n(n-1)\ldots(n-k+1)}{n^k} = \left(1-\frac{1}{n}\right)\left(1-\frac{2}{n}\right)\ldots\left(1-\frac{k-1}{n}\right)$$

$$\geq 1 - \frac{1+\cdots+k-1}{n} = 1 - \frac{k(k-1)}{2n}.$$

Thus, for any A,

$$|M_{uk}(A) - H_{uk}(A)| \leq \frac{k(k-1)}{2n}. \qquad \text{QED}$$

Remark: Note that the preceding calculations simply involve calculating the chance that a sample of size k from $\{1, 2, \ldots, n\}$ with replacement yields all distinct elements. This is just the birthday problem of chapter 1. It is straightforward to show that for the chosen A, $|M_{uk}(A) - H_{uk}(A)| \geq 1 - e^{-k(k-1)/2n}$. Thus our analysis is sharp; k^2/n must be small to have the two distributions close.

To conclude, let P be an exchangeable probability on sequences of length n. We saw that P can be exactly represented as a mixture of urn measures H_u. The mixing measure μ is simply induced by P: pick from P, what's the chance that the induced sample gives the urn u? Now,

$$|P_k(A) - M_{\mu k}(A)| = \left|\int H_{uk}(A)\,\mu(du) - \int M_{uk}(A)\,\mu(du)\right|$$

$$\leq \int |H_{uk}(A) - M_{uk}(A)|\,\mu(du) \leq \frac{k(k-1)}{2n}.$$

This concludes the proof of the general de Finetti theorem.

Remark: Of course, the version of de Finetti's theorem for arbitrary $\mathcal{X}$ yields a theorem when $\mathcal{X} = \{0, 1\}$. For this case, something sharper can be said. By using a sharper bound for sampling with and without replacement for urns with only two kinds of balls, Diaconis and Freedman proved the following theorem.

Theorem: Let P be an exchangeable probability on binary sequences of length n. Then there exists a probability μ on $[0, 1]$ such that

$$|P_k(A) - P_{\mu k}(A)| \leq \frac{4k}{n} \qquad \text{for any set } A.$$

Thus only k/n needs to be small (rather than k^2/n being small).

FROM FINITE TO INFINITE

The last step in deriving the usual version of de Finetti's theorem for infinite exchangeable sequences, as before, requires taking limits in the theorem of the last section. Let P be an exchangeable probability on infinite sequences of elements of a set $\mathcal{X}$. Thus for any n and sets $A_1, A_2, \ldots, A_n$ of $\mathcal{X}$, $P(A_1, A_2, \ldots, A_n)$ is defined, exchangeable, and extendable. From the main theorem of the last section, there exists a mixing measure, call it μ_n (because it depends on n), such that for any $k \leq n$,

$$|P(A_1, \ldots, A_k) - P_{\mu_n k}| \leq \frac{k(k-1)}{2n}.$$

Fixing k, let n tend to infinity. The μ_ns have a limit μ in an appropriate sense, so

$$P(A_1, \ldots, A_k) = P_{\mu k}.$$

Since this holds for all k with the same μ, $P = P_\mu$ follows.

The existence of the limit μ_n in the space of all probabilities $\mathcal{P}$ is a classical part of functional analysis. Details are in Diaconis and Freedman (Section 3). This limit is what forces mild restrictions on $\mathcal{X}$. Indeed Dubins and Freedman* show that $P = P_\mu$ can fail for completely general $\mathcal{X}$. The failure leans on exotic set-theoretic notions such as the axiom of choice and nonmeasurable sets. For any concievable real-world situation, de Finetti's theorem holds.

There is a finitely additive version of the theorem which holds for any $\mathcal{X}$. See Diaconis and Freedman (Theorem 29).

---

*L. Dubins and D. Freedman, "Exchangeable Sequences Need not Be Mixtures of Independent, Identically Distributed Random Variables," *Zeitschrift für Wahrscheinlichkeits theorie und verwandte Gebiete* 48 (1979): 115–32.

Per Martin-Löf

CHAPTER 8

——

ALGORITHMIC RANDOMNESS

Can computers generate random sequences? Random number–generating programs purport to do so. Can nature generate random sequences? Do we really have a clear idea what an objectively random sequence really is? You will remember that when we left von Mises theory in chapter 4, even at that highly idealized level, the theory of a random sequence was not theoretically satisfactory.

But random sequences are also important at a very practical level. In almost every walk of scientific life, simulations have a role to play, and these require sources of random numbers. So do secret codes used by spies, banks, and the Internet for secure communications and secure transactions. It turns out that often they are not sufficiently random, and then some bad science and bad security results.

The efforts to generate and test random numbers have deep philosophical tendrils. Our chapter begins at the practical end and then turns to logicians' efforts to define perfect randomness. The last section combines these concerns.

COMPUTER GENERATION OF RANDOM NUMBERS

Computers usually work with finite things, and we focus on an easy, widely used case: generating random numbers in the natural numbers $\{0, 1, 2, \ldots, N\}$. Thus one wants a sequence, $X(1), X(2), X(3), \ldots$, in this range that is *uniform* and *independent*.

The standard schemes are deterministic. They choose $X(1)$ "somehow"—by human input of the seed or using the number of milliseconds since the start of day—and proceed by

$$X(n + 1) = f(X(n)),$$

with f being a fixed function. A classic example, RANDU, had $f(j) = 65{,}539 \cdot j \pmod{2^{31}}$. RANDU was a widely used random number generator in the 1960s, until it was shown that when used to plot random points in three dimensions—the coordinates of a point being three consecutive "random" numbers—the points clustered on planes. This is a decidedly nonrandom outcome! In 1968 George Marsaglia published a proof that all schemes in the same general class as RANDU would have the same defect.[1]

More sophisticated variants use higher order recursions. In an early "bible" of random number generation,[2] the author's favorite generator was $X_{n+1} = X_{n-24} \cdot X_{n-55} \pmod{2^{32} - 1}$. This requires starting with seed $X_1, X_2, \ldots, X_{55}$. The most popular modern generator, Mersenne Twister,[3] uses a similar scheme.

Do these schemes work? Well, yes and no. For some tasks, such as computing integrals and playing computer games, they have usually done well. However, there is a long history of failures as well. In 1993, a New York Times article[4] recorded a failure of several new and improved random number generation schemes to solve routine instances of a statistical physics problem whose correct solution was known analytically. We know casino hustlers who capitalize on the fact that today's slot machines work with the very simple generators just described. By observing a few hundred pulls of the handle, they figure out N and $f(i)$. If they know these and the current $X(n)$, they know $X(n+1)$, $X(n+2)$, and so on. They watch play until a big jackpot is due. Then they "accidentally" spill coffee on the current player and (after Oops, please take this $50 chip for cleaning), they take over, play, and collect. Bank fraud and security break-ins on computer systems are reported daily.

If casinos, bankers, and the CIA can't get it right, this suggests that there might be a basic problem! Random numbers are usually tested by using a battery of ad hoc tests. (For instance, will high and low follow each other like coin tossing should? What about consecutive sets of 3, odd, and even, and so on?) These tests embody good common sense, but we must remember that random numbers are called on for very disparate tasks. Good for some tests does not mean good for others.

For Donald Knuth's random generator, the test that does it in is called the birthday spacings test. Get the generator to generate $X(1)$, $X(2)$, ..., $X(500)$, between 1 and 1,000,000. Order these (say, from smallest to largest). Consider the "spacings," largest–second largest, second largest–third largest, ... Look at the number of repeated values among the spacings. The approximate distribution for the number of repeats can be determined theoretically, and Knuth's generator produced numbers 16 times too large. There are some tests with some claim to generality: if you pass this test, you will pass a whole slew of other tests. (One is called the *spectral test*.) But these are far too limited to capture the vast range of applications.

It is natural to ask for Nature's help. After all, quantum mechanics and thermal noise are supposed to be truly random. We consulted with a wonderful group of physicists who wanted to put a huge supply of good random bits on a disk at the end of their book.[5] Here is what they wound up doing. They started with "electrical noise" measured via the delay times in a leaky capacitor. This generated a long series of random binary digits. Testing showed that these were not so random. You could see periodic fluctuations, which were traced to the 24-hour variations in the power supply! A thousand such strings were generated. They were combined into one string by taking the sum (mod 2) of the thousand binary digits in each position. This final string was then scrambled using the data encryption standard. These were the final digits used.

A host of other techniques have been proposed. These range from using the quantum mechanical fluctuations in the clicks of a Geiger counter to using a lava lamp. Such schemes can be combined with each other and with the deterministic mathematical generators described before. There are no theoretical guarantees, and it all seems terribly ad hoc for such an important enterprise.

One hopeful development is the use of the logic of complexity theory, as in the scheme of Blum and Michali.[6] These authors offer a generator with the following property: if it fails any polynomial time test, then there is an explicit way to factor that is much faster than any known method. (Thus, if factoring is hard, our numbers are secure. But, of course, if factoring can be done efficiently, all bets are off.) This is close in spirit to the algorithmic complexity accounts treated subsequently in this chapter.

We conclude this practical section with some practical advice on the use of random number generators:

> Use at least two generators and compare results. (We recommend Mersenne Twister and one of the generators offered in *Numerical Recipes*.)
>
> Put in a problem with a theoretically known answer to run along with the other simulations.

Think of using random number generators like driving a car. Done with care, it is relatively safe and useful.

ALGORITHMIC RANDOMNESS

The algorithmic theory of randomness, our eighth great idea, reached its modern form with Per Martin-Löf's 1966 "The Definition of Random Sequences."[7] The concept of an objectively random sequence—one that is perfectly random—is made precise using the theory of computation.

As we saw in chapter 4, the problem that the algorithmic theory of randomness solves was put forward by Richard von Mises in 1919. Von Mises wanted to ground the application of probability to reality by putting forward an idealized mathematical model of random phenomena. We can see this as an attempt to sidestep the fallacies that small-probability events don't happen. Instead of starting with the random process of coin flipping and arguing that it is "morally certain" that the resulting random sequence would have certain properties, von Mises wants to give the theory of a random sequence directly.

Let's recall where we left his program: (1) A random sequence of 0s and 1s should have a limiting relative frequency, and (2) limiting relative frequency should be the same for any infinite subsequence selected out by an admissible place-selection function. (Admissible is left to be defined.) For example, consider the sequence of alternating 1s and 0s:

$$101010101010101010101010101010101010 \ldots .$$

The limiting relative frequency of 1s is $\frac{1}{2}$. But the place-selection function that selects odd members of the sequence gives

$$11111111111111111111111111111111111 \ldots ,$$

and that which selects even members gives

$$000000000000000000000000000000000000s \ldots$$

for limiting relative frequencies of 1 and 0, respectively.

Do von Mises random sequences exist? That depends on the class of admissible place-selection functions. Too few place-selection functions give patently nonrandom sequences. For an extreme example, suppose we have just the preceding two place-selection functions. Then the sequence

$$11001100110011001100 \ldots$$

would count as random, since each of the two place-selection functions selects a sequence

$$1010101010101010 \ldots ,$$

which has a limiting frequency $\frac{1}{2}$, just as the original sequence. (Notice that in these examples, the limiting relative frequency of 1s is approached from above, e.g., 1. $\frac{1}{2}, \frac{2}{3}, \frac{1}{2}, \frac{3}{5}, \frac{1}{2}, \frac{4}{7}, \ldots$)

But you can easily think of an additional place-selection function that will change the relative frequency, for instance, pick every fourth entry. And you could then easily construct a sequence that is random according to the expanded class of place selections.

This raises a general question. What if there are a lot of place-selection functions? Can we always cook up a sequence that is random according to all of them? The answer depends on just what you mean by "a lot."

Critics of von Mises[8] were quick to point out that if *all* functions, in the set-theoretic sense, are included, then there are simply no random sequences. Abraham Wald[9] (who thought of functions not as sets but rather as rules that could be explicitly described) proved that given any countably infinite set of place-selection functions, one can indeed cook up von Mises random sequences relative to that class of place-selection functions. The question was left hanging as to whether there was some natural set of functions to use.

An answer was suggested by the theory of computability, but this was possible only after that theory had been developed by Turing, Gödel, Church, and Kleene in the 1930s. The idea of applying computability to von Mises' definition of a random sequence was due to Alonzo Church in 1940.[10]

Church suggested taking the admissible place selection functions as the computable ones—ones that could be implemented by a Turing machine.

This seemed like a natural choice, and since there are a countable number of Turing machines and a noncountable number of sequences, there are plenty of sequences that are random in this sense.

Unfortunately, the definition has a flaw. Von Mises–Church random sequences lack some of the properties that they should have. In particular, some sequences that are random in this sense approach their limiting relative frequency from one side. This means that they are vulnerable to a gambling strategy. Von Mises had considered the impossibility of a successful gambling system the sine qua non of a random sequence:

> By generalizing the experience of the gambling banks, deducing from it the Principle of the Impossibility of a Gambling System,

and including this principle in the foundation of the theory of probability, we proceed in the same way as the physicists did in the case of the (conservation of) energy principle.[11]

Such violating sequences could hardly be taken as paradigms for outcomes of fair coin flips.

In a deeper sense, Church's idea of using computability to define randomness was correct. The problem was the way in which it was used. In fact, the vulnerability to betting systems has nothing to do with computability. Ville, in 1939,[12] showed that for *any* countable set of place-selection functions, there is a von Mises random sequence in which relative frequency of H is $\frac{1}{2}$, but for all but a finite number of initial segments, the relative frequency of H is not less than $\frac{1}{2}$. (Like our previous examples: 10101010 . . . and 110011001100. . . . In fact, the random sequences constructed by Wald to prove the consistency of von Mises' definition all had this property.) The set of place-selection functions proposed by Church is countable, so it inherits the problem.

The source of the problem is not the idea of using computability. Rather, von Mises' idea of defining randomness by means of place-selection functions *alone* appears to be defective.

Church–von Mises randomness requires random sequences to pass only one sort of test for randomness. Sequences can pass this test and fail others. We want random sequences to pass all tests of randomness, with tests being computationally implemented.

Per Martin-Löf found how to do this in 1966.[13] We shall see that two other, apparently different, ways of incorporating computability in a definition of randomness gave definitions that turned out equivalent to that given by Martin-Löf.

COMPUTABILITY

If you know how to program in any computer language, you already know what computability is. They all allow you to compute the same functions, although the required program may be shorter in some languages than others. Nevertheless, we include a brief account of the birth of computability theory because it provides a case study in the robust mathematical explication of a philosophical notion.

What is a computation? It is a philosophical question that goes back at least to Hobbes and Leibniz. Hobbes maintained that all thought was a kind of calculation—an idea that had a second life in the artificial intelligence community in the late twentieth century. Leibniz envisioned a universal formal language of thought and a system of rules of valid inferences, which together reduce any truth step by step to an identity. Empirical truths would require an infinite number of steps, recapitulating God's reasoning in deciding whether to create this world rather than another. But mathematical truths required only a finite number of steps, so in principle any mathematical question could be settled by logical analysis. To this end Leibniz worked both on formal logic and on the invention and construction of a calculating machine.

Leibniz' program, sans theology, lived into the twentieth century in the views of Bertrand Russell and David Hilbert. Russell held that all mathematics was reducible to logic. Hilbert thought that every mathematical problem could be decided. Hilbert put a sharp-enough point on the problem of computation to motivate its solution, in his statement of the *Entscheidungsproblem*: "Is there an algorithm that takes as input a mathematical statement and outputs 1 if it is true and 0 if it is not?" Hilbert and Ackermann (in 1928)[14] specifically ask the question for first-order logic (logic with individuals, predicates, and quantifiers *all* and *some* over individual variables): "The Entscheidungsproblem is solved when we know a procedure that allows for any given logical expression to decide by finitely many operations its validity or satisfiability." Using ideas pioneered by Kurt Gödel, a negative answer was found almost simultaneously by Alonzo Church and by Alan Turing. Leibniz was wrong!

The analysis required, at the onset, a precise theory of computation. A number of rather different ideas were put forward.

TURING COMPUTABILITY

Computers, at the beginning of the twentieth century, were people. They sat at desks and carried out computations according to instructions, using paper and pencil. Alan Turing gave an abstract version of this process allowing unlimited paper and thus produced the very simple and intuitive notion of a Turing machine. Think of the pieces of paper available for computation strung together as a tape, infinite in both directions. Each piece of paper is called a cell of the tape. At any time, the head of the machine is over some cell, scanning what is written in the cell—which can either be a 0, 1 or B (blank). There is a special cell designated as the starting cell. There is a finite set of internal states, including a special starting state. Depending on the symbol scanned and the internal state of the machine, the machine performs an operation, either

Write 0, 1, or B in the cell it is scanning.
Move one cell to the left or to the right.

Alan Turing

When the action is taken, the machine takes on a (possibly) new state. Thus the dynamics of the machine in discrete time is characterized by a set of quadruples:

<current state, symbol scanned, operation, new state>,

There are no two quadruples with the same initial pair; the dynamics is deterministic. If the machine is in a state, scanning a symbol, such that there is no corresponding quadruple in its instruction set, the machine *halts*.

Here is an example. The machine starts with an empty tape, full of blanks. It is in its start state, which we will call S_0. It prints a zero and enters state S_1. This is accomplished by the quadruple

$$<S_0, B, 0, S_1>.$$

Now it is in state S_1, scanning the 0 that it just printed. The following instruction tells it to move to the right one cell and assume a new state, S_2:

$$<S_1, 0, R, S_2>.$$

Now it is again scanning a blank, but in state S_2. The next instruction tells it to print a 1 in that blank and assume state 3:

$$<S_2, B, 1, S_3>.$$

It is now in state S_3, scanning a 1. Our final instruction tells it to move to the right and go back to state S_0.

$$<S_3, 1, R, S_0>.$$

It is now in state S_0, scanning a blank, just as it started; these four instructions define a Turing machine that prints out the infinite sequence

$$01010101010101. \ldots$$

Everything is finite about the machine (states, symbols, and instruction set), with the exception of unlimited tape. For a machine that computes the value of a function, the tape is assumed to start with the head over the leftmost cell of encoded values of the arguments. The machine halts with the head over the encoded value of a function. The machine may not halt for certain inputs, and in this case the function is a partial function.

For validity of first-order logic to be *decidable*, there would have to be a Turing machine that, when input a suitably encoded version of a formula of first-order logic, would output a 1 if the formula is valid and a 0 if not. A (nonempty) set is *computably enumerable* if it is the range of a total computable function. That is to say that there is a Turing machine that when given inputs 1, 2, 3, . . . would list the members of the set. Although Gödel showed (by his completeness theorem) that the valid formulas of first-order logic are computably enumerable, Turing and Church showed that that validity is not decidable.[15] The *Entscheidungsproblem* for first-order logic is unsolvable.

Turing did this via another unsolvability result. First, he showed that it was possible to build a *universal Turing machine*—one that could emulate every other Turing machine if fed the appropriate input. Then he asked whether there was a Turing machine that decides the halting problem for all Turing machines. Is there, that is to say, a machine that when fed in the description of an arbitrary machine and an input for that machine outputs a 1 if the target machine will halt with that input and a 0 if not? He showed that the supposition that there is a machine that decides the halting problem leads to a contradiction. If there were such a machine, it would be possible to construct another that halts if and only if it doesn't. This undecidability of the halting problem leads to a proof of the undecidability of the decision problem for first-order logic.

RECURSIVE FUNCTIONS

Church took a different route to computability. His first approach was through his λ-calculus, which is the basis for the programming language LISP. Later he and his student Stephen Kleene, following the lead of Gödel, used recursive functions.

The *primitive recursive functions* are gotten from *zero, successor, and projection* functions, by closure under *composition* of functions, and *primitive recursion*:

The *zero* functions are $f(x_1, \ldots, x_k) = 0$.

Successor $S(x) = x + 1$,

Projection $I_n^k(x_1, \ldots, x_k) = x_n$.

Composition of f and $g_1, \ldots, g_n$, (bold **x** is a vector):

$$h(\mathbf{x}) = f(g_1(\mathbf{x}), \ldots, g_n(\mathbf{x})).$$

Primitive recursion from f and g:

$$h(\mathbf{x}, 0) = f(\mathbf{x}),$$

$$h(\mathbf{x}, t+1) = g(t, h(\mathbf{x}, t), \mathbf{x}).$$

The *general recursive partial functions* are gotten from the primitive recursive functions by closing under a minimization operator, μ, that gives the smallest argument that gives the function a value of 0, if there is one. Where there isn't, the function is undefined. That is,

$$h(\mathbf{x}) = \mu y(g(\mathbf{x}, y) = 0)$$

is the smallest solution to the equation $g(\mathbf{x}, y) = 0$ if there is one and is undefined for values $\mathbf{x}$ where there is no smallest solution.

Turing proved that the general recursive partial functions are the Turing computable functions. They are undefined for inputs for which the machine doesn't halt. Two rather different ideas lead to the same place.

PROGRAMING LANGUAGES AGAIN

General recursive functions are also the functions that can be programmed in any modern computer language (size restrictions removed). If the programming language is restricted so that only bounded loops are possible, then it can compute only primitive recursive functions. With unbounded loops we get only partial functions because the program may not halt.

A ROBUST EXPLICATION OF COMPUTABILITY

All sorts of other approaches, including Markov algorithms,[16] Church's λ-calculus, much fancier Turing machines and register machines, computers with random access memory, and so forth, have been shown to lead to the same class of computable functions. This lends confidence to the intuition that this is the "right" notion of computability. In 1946 Kurt Gödel wrote, ". . . with this concept one has for the first time succeeded in giving an absolute notion to an interesting epistemological notion, i.e., one not depending on the formalism chosen."[17]

Kurt Gödel

RANDOMNESS

Since Church–von Mises randomness proved inadequate, Martin-Löf took a different approach. One could define random sequences by throwing out "atypical" classes—*null* classes—that were given probability 0 by a model of flipping a fair coin. The immediate problem is that each individual sequence has probability 0. Here computability comes to the rescue. We throw out only null sets that can be identified by a Turing machine. Since there are a countable number of Turing machines, throwing out all these null sets leaves a set of "typical" random sequences with probability measure one. (This again can be regarded as a way of avoiding the absurdities, discussed in chapter 4, of a literal application of Cournot's principle.)

The nonrandom sequences can be thought of as those failing more and more stringent statistical tests. Suppose we see reports of coin tossing:

010101.

It is a little suspicious. Suppose the sequence continues:

01010101010101010101010.

That is a lot more suspicious. It is a lot less likely that fair coin flips should produce this sequence. (All the practical tests of computer programs to generate random numbers, with which we began this chapter, are of this character. They ask, in different ways, how likely it is that a sequence of coin flips could have produced this sequence.) Martin-Löf tests of randomness are a computable implementation of increasingly stringent tests of randomness. Here is how it goes.

Consider the set of infinite sequences of zeros and ones that continues some finite initial segment. These are called cylinder sets. Considered as sequences of tosses of a fair coin, any set corresponding to an initial segment of length n has probability $\left(\frac{1}{2}\right)^n$. A countable union of disjoint cylinder sets has a probability that is the sum of the probabilities of those sets.

Martin-Löf uses computability twice:

1. We restrict our attention to the case of unions of *computably enumerable* sequences of cylinder sets. That is, one can have a Turing machine that prints out characterizations of these, one after another. Call these the *effective sets*.

2. Less and less likely effective sets are intuitively less and less typical of a random sequence. One hundred alternations of heads and tails are quite suspicious. We can use a nested sequence of such sets that are less and less probable to approximate a definitely nonrandom set. That is, we consider a sequence $U_1, \ldots$ such that U_{n+1} is a subset of U_n, and the probability of U_n is at most $\left(\frac{1}{2}\right)^n$. We require that this sequence be *computably enumerable*. Such a sequence is a *Martin-Löf test*. It enumerates more and more stringent tests.

A set that is in the intersection of such a sequence is a *constructive nullset*. It fails the Martin-Löf test of randomness. A Martin-Löf random sequence is then defined as one that passes *all* the tests—one that is in no constructive null set.[18]

We can illustrate this by showing how our infinite sequence of alternating 0s and 1s fails to be random. The reasoning applies equally well to any computably enumerable sequence. Consider a cylinder set determined by the first n members of the sequence. This is an especially simple *effective set*.

The sequence of all cylinder sets determined by longer and longer initial segments of our original sequence:

all continuations of 0,
all continuations of 01,
all continuations of 010,
all continuations of 0101,

and so on, is thus a Martin-Löf test. The intersection of these cylinder sets is a Martin-Löf nullset whose only member is our original sequence, so it is not random.

There are a countable number of constructive nullsets, and each has probability zero, so the probability that a sequence is Martin-Löf random is equal to 1. Martin-Löf shows that each sequence that is random in his sense has a limiting relative frequency—a property that von Mises had to postulate separately for Kollektivs. This extends and unifies the connection between relative frequency and chance.

COMPUTABLE MARTINGALES

We have seen that von Mises considered the impossibility of a successful gambling system essential to the definition of a random sequence. This was his justification of the use of place-selection functions in his definition of randomness. You should not be able to make fair bets on a subsequence of a random sequence in a systematic way and make money. But Jean Ville had shown that the definition in terms of place selection is too restrictive. Some sequences that are random in the sense of Mises-Church are vulnerable to a gambling system—but not one so simple as to rely on computable place selection.

The idea of a gambling system for a sequence is made precise in the notion of a *martingale*. Suppose that you start with unit capital. You put some proportion of your capital in a bet that the first member of the sequence is a 1 or 0. If you stake everything on 1 and it is 1, you win and your capital is now doubled. Now you proceed, allocating proportions of your current stake on bets on the next member according to a rule that can consider the whole history of the sequence up to the point of the bet. The gambling strategy succeeds against a sequence if your stake grows to infinity.

Instead of explicitly writing down the strategy, the martingale can be characterized by the stake in hand at every point in the sequence.

So viewed, a martingale is a function, CAP, from initial segments to nonnegative reals, such that

$$\text{CAP}(s) = \tfrac{1}{2}\,[\text{CAP}(s \text{ followed by } 0) + \text{CAP}(s \text{ followed by } 1)]$$

since the odds are fair. A martingale *succeeds* on a sequence if CAP goes to infinity in the limit.

For example, consider a betting strategy that puts all the capital on 1 if the last event is 0 and all the capital on 0 if the last event is 1. The resulting capital function is a martingale. It succeeds on the sequence

$$010101. \ldots$$

It does not succeed on 00110011. . . . The strategy immediately goes bust on this sequence; CAP(00) = 0. But another martingale will succeed on this sequence.

It remains to impose computability constraints. The martingale is required to be computably enumerable. (The martingale is, in general, a function to real numbers, so computability is imposed by approximation. It is said to be computably enumerable if it is *left-computably enumerable*. That is to say, it is approximable from below by rational-valued functions.) Then one can define a random sequence in terms of impossibility of a gambling system. A sequence is random if no computably enumerable sequence succeeds in it. In 1971 Schnorr[19] proved that a sequence is random in this sense if and only if it is Martin-Löf random.

KOLMOGOROV COMPLEXITY

In the 1960s Kolmogorov returned to the foundations of probability from a new standpoint.[20] A finite sequence of 1000 alternating 0s and 1s is *algorithmically compressible* in that one can write a short program to generate it. This is because of its simple structure. Absence of structure is a way to define randomness. One can then measure randomness of a finite sequence, relative to some universal Turing machine, by measuring the length of the shortest program that will generate it. The shorter the program, the less random the sequence. Kolmogorov complexity was introduced independently by Gregory Chaitin in 1966.[21]

But the degree of randomness, thus defined, depends on the universal Turing machine (or the programming language) chosen. The dependence of the result on the universal Turing machine is limited, however, since any such machine can be programmed to simulate any other. The length of the simulation program is a constant, and length of shortest programs to generate a sequence must agree up to this constant. The differences may wash out at infinity (if everything goes well), but plausible application to finite sequences appears to depend on a plausible natural choice of a universal Turing machine.

In this approach to randomness or computational complexity, Kolmogorov and Chaitin were anticipated by Ray Solomonoff.[22] Solomonoff was partly inspired by a class on inductive logic taught by the philosopher Rudolf Carnap, which he sat in as an undergraduate at the University of Chicago. He thought that Carnap's project was good but that he was going about it in the wrong way. Solomonoff's idea was to use computational complexity in the construction of a universal prior.*

Since computational complexity depends on the choice of a Turing machine, or programing language, used to measure it, so does the universal prior. Some discussions try to slide by this by noting that one universal Turing machine can emulate any other, given the appropriate code. But the appropriate code may be long, and the difference in complexity may be large. Solomonoff, however, saw this not as a problem to be minimized, but just as a fact of life. The choice is subjective; it is a judgment based on our general experience. It is the choice of a prior, which is then updated by experimental results.

Solomonoff is unabashedly a subjective Bayesian:

Subjectivity in science has usually been regarded as Evil—that is something that does not occur in "true science" that if it does occur, the results are not "science" at all. The great statistician, R. A. Fisher, was of this opinion. He wanted to make statistics a "true science" free of all the subjectivity that had been so much a part of its history.

I feel that Fisher was seriously wrong in this matter. . . .

---

* The idea is spectacularly successful, provided that God uses a Turing machine to generate the universe.

In ALP (algorithmic probability), this subjectivity occurs in the choice of "reference"—a universal computer or universal computer language.[23]

Ray Solomonoff

It seems natural to apply ideas of computational complexity to von Mises' problem of defining a Kollektiv by passing to the limit. A Kollektiv would be an infinite sequence with incompressible initial segments. Applying computational complexity, as introduced above—*plain Kolmogorov* computational complexity—infinite random sequences do not exist!

Although the initial attempt to define a random infinite sequence in terms of Kolmogorov complexity failed, this turned out to be because the issue was not framed in exactly the right way. That is because a program (input) for a universal Turing machine contains both information about the sequence to be computed and also information about the length of the program. For the algorithmic complexity of the sequence, we are interested only in information about the sequence being computed. This can be achieved by restricting attention to a special kind of universal Turing machine—a *prefix-free* universal Turing machine. We measure the prefix-free complexity of a sequence K using a prefix-free universal Turing machine.[24, 25]

Then, at infinity, everything works. An infinite sequence is algorithmically random if there is a constant c such that complexity K of every initial segment of length n is greater than or equal to $n - c$. Algorithmically random infinite sequences now exist, and Schnorr (1971) also proved that these random sequences also coincide with the Martin-Löf random sequences.

There is thus one very strong and reasonably robust concept of algorithmic randomness. As Martin-Löf remarked, it gives a correct definition of von-Mises' idea of a Kollektiv. The frequentist ideas of von Mises are not so opposed to the measure-theoretic framework of Kolmogorov after all. *Flipping a fair coin infinitely often produces a Kollektiv with probability 1.*

VARIATIONS ON RANDOMNESS

If randomness is based on computation, then variations on randomness can be based on different computational notions. The theory can be pushed into higher realms of abstraction by equipping a Turing machine with an oracle. An oracle is a black box that will answer any of a certain class of questions that the Turing machine can put to it at any point in a computation. One can consider a universal Turing machine equipped with an oracle that decides the halting problem for any Turing machine. Such a machine is computationally more powerful than any Turing machine. Of course, such a machine does not decide the halting problem for Turing machines with such an oracle because that would lead to a contradiction.

One can then consider a super-oracle that decides the halting problem for these machines, and so forth, creating a hierarchy of computationally more powerful machines. This leads to a hierarchy of more and more stringent criteria for a random sequences, with Martin-Löf randomness being 1-randomness, the same notion with a Turing machine equipped with an oracle deciding the halting problem being 2-randomness, and so on. The class of $N + 1$ random sequences is strictly contained in the class of N-random sequences.

In the other direction, weaker kinds of randomness can be gotten by limiting the kind of computation used to test for randomness.

This can be done in various ways. Schnorr proposed computable tests rather than computably enumerable tests, which gives a weaker notion of randomness. There is a theory of *P-randomness* using computability in polynomial time that parallels the theory discussed above. P-random sequences still possess many of the features one would want from a random sequence. Limiting computational resources in various ways gives weaker notions of randomness. If we are willing to fix on a programing language that we find natural, then Kolmogorov complexity gives us a usable measure for finite sequences.

SUMMING UP

The quest for the notion of an objectively random sequence had its origins in von Mises' attempt to formulate a pure frequency theory of probability based on an objectively disordered random sequence—his Kollektiv. This, most of us would agree, is a failed foundational program. The modern theory of algorithmic randomness proceeds within the Kolmogorov framework rather than being an alternative.

Now that Martin-Löf and others have developed a satisfactory concept of such a sequence via computability theory, we can ask what remains of von Mises' program. Returning to Borel (chapter 5), we can now say that Bernoulli trials—coin flipping—produce a von Mises Kollektiv with probability one. Then, due to de Finetti (chapter 7), exchangeable degrees of belief are equivalent to belief in a von Mises Kollektiv with uncertain frequency.

Democritus

CHAPTER 9

PHYSICAL CHANCE

Here I saw both Socrates and Plato . . . Democritus,
who sets the world at chance; Diogenes, Anaxagoras,
and Thales, Empedocles, Heraclitus, and Zeno

—DANTE *INFERNO* CANTO IV

Our ninth great idea is that the world is fundamentally a world of chance. This is true even in classical physics, due to sensitive dependence on initial conditions and resulting chaotic behavior. It is all the more true when we move to quantum mechanics. This chapter explores these ideas, and shows how all the probabilities involved can be thought of in the same way.

In 1731 Daniel Bernoulli, whom we met in connection with the St. Petersburg problem, pioneered the foundations of the kinetic theory of gases. Gases were composed of particles in motion, with pressure on a container being the result of collisions of the particles and temperature being determined by their kinetic energy. The basic idea of air being composed of particles too small to observe goes back to the Greek atomists Empedocles and Democritus. Lucretius even uses the irregular motion of dust motes in sunbeams as evidence of their collisions with smaller particles:

> Or you might say that it is worthwhile to study
> the way in which the motes of dust dance in a sunbeam
> because the behavior of these tiny objects
> gives us a notion of that of invisible particles.
> You will see many of these sailing dust-motes
> impelled no doubt by collisions one cannot detect,
> change direction, and turn off this way or that.
> Surely their movement depends on that of the particles.

—Lucretius *De Rerum Natura* Book II

But Bernoulli was able to turn these brilliant speculations into science. He derived an ideal gas law: for constant temperature, pressure is inversely proportional to volume. And he predicted deviations from this law for dense gases. Although he is dealing with a chance process and is certainly familiar with probability, probability theory does not play a part in his analysis. This is one of the fundamental tensions of statistical mechanics: In trying to understand a complicated *deterministic* system, probability is introduced, calculations are done, predictions are made, and it works. The predictions are non-probabilistic. So what is probability doing in the middle of all this?

BOLTZMANN

Probability theory enters, somewhat reluctantly, in the work of Ludwig Boltzmann. Boltzmann pursued the explanation of Clausius' thermodynamics—in particular, the second law that entropy increases—by the atomistic hypothesis. The setting is a dilute gas in a box, with its constituents modeled as balls undergoing elastic collisions with each other and the walls of the container. Originally, the second law was to be given a strict demonstration on the basis of mechanics alone. Boltzmann (1872)[1] assumes a hypothesis of molecular chaos (the *Stosszahlensatz*) earlier introduced by Maxwell (1867). On this assumption, Boltzmann proves his *H*-theorem. His *H* function decreases monotonically in time and remains constant when the gas reaches Maxwell's equilibrium distribution. Entropy is $-kH$. Boltzmann concludes (Quoted in Jos Uffink, "Compendium to the Foundations of Classical Statistical Mechanics," in *Handbook for the Philosophy of Physics*, ed. J. Butterfield and J. Earman (Amsterdam: Elsevier, 1074): 924–1074)

   This provides an analytical proof of the Second Law . . . .

This claim was soon subjected to two devastating challenges. The first was raised by Josef Loschmidt in 1876[2] in the context of a more complicated debate. Loschmidt's general point was that since the laws of Newtonian mechanics are time-reversal invariant, it is impossible to have a proof that entropy must increase on the basis of those laws alone. In the context at issue, if the system can evolve from a low-entropy state at time t_1 to a high entropy state at time t_2, then reversing the velocities of the particles at t_2 will lead the system to evolve from a high-entropy state to a low-entropy state.

In 1893 Loschmidt's *reversibility objection* to the supposed derivation of the second law was joined by another, the *recurrence objection*. Ernst Zermelo, in 1896[3] using earlier results of Henri Poincaré,[4] showed in a precise way that monotonic increase in entropy was impossible. For almost all* initial states, the system must eventually return (arbitrarily closely) to the initial state. Since entropy is a continuous function on phase space, it must return arbitrarily close to its original value.

But Boltzmann had *proven* the *H*-theorem. So the objections had, for some time, an air of paradox. Subsequent analysis showed that the temporal asymmetry of the *H*-theorem resulted from a temporal

---

* That is, except for a set of probability 0, where probability is given by uniform measure over all states with the same energy.

asymmetry in the formulation of the *hypothesis of molecular chaos*. This hypothesis, which appeared uncontroversial to Maxwell and Boltzmann, makes an independence assumption prior to a collision of gas molecules. It is not equivalent to its time reversal. If the assumption is made postcollision, then it can be proved that entropy monotonically decreases. Both the hypothesis and its time reversal are true only when the gas is in equilibrium.

At the end of his discussion, Zermelo concluded that either the second law—as originally formulated—must be given up or the kinetic theory of gases—in its contemporary form—must be given up. Poincaré was already an opponent of the kinetic theory and used this as an argument against it. Boltzmann modified his view of the second law. In his replies to Loschmidt and to Zermelo, he says that the second law has only a statistical character.

The basic idea is that many microstates correspond to the same (macroscopic) observable state. Entropy is defined according to the partition of microstates induced by these observable macrostates. High-entropy macrostates correspond to many more microstates than low-entropy ones and are consequently held to have higher probability. Boltzmann now holds that if the system is in a low-entropy state, it is very probable that it will move to a higher-entropy state. He also holds that a low-entropy state is very likely to have come from a high-entropy state, so time symmetry is respected while some of the flavor of the second law is retained. This new probabilistic view is quite different from the traditional second law, which is fundamentally time asymmetric.

Let us consider this more closely. Suppose that a system starts in a low-entropy macrostate. What is the argument that it will transition to a higher one with high probability? If the transition consisted in drawing a new microstate at random from the set of all microstates, then the result would follow.* But the transition is not determined in this way but is rather determined by the dynamics of the molecules of gas. The new statistical argument depends upon a promissory note about the nature of the dynamics.

---

*Technicality: "At random" is being cashed in using a uniform probability on the constant-energy hypersurface.

The nature of the probabilistic view of the second law was substantially clarified by Paul and Tatiana Ehrenfest in 1911 in an article written for an encyclopedia of mathematical sciences.[5] Paul Ehrenfest was Boltzmann's student. The article was initially supposed to be written by Boltzmann himself, but the task fell to the Ehrenfests after Boltzmann's suicide in 1906.

In the article, now translated as *The Conceptual Foundations of the Statistical Approach in Mechanics*, the Ehrenfests analyze a simple model in terms of which many of the conceptual questions become transparent. Think about two dogs standing near one another; one has a lot of fleas and the other has none. Each flea has an identity. Every moment one flea, chosen at random, jumps from one dog to the other. It is intuitively obvious that if there are a lot of fleas, then there is a tendency to equalize.

One can think, less fancifully, of two urns with a lot of numbered balls distributed between them. At each time step, a ball is selected at random and moved to the other urn. This is known as the Ehrenfest urn model. Which balls are in which urn fix the microstate state of the system; how many balls are in each urn fix the macrostate. The transition probabilities are known, and the model can be explicitly solved.

Typically, we consider a large number of balls, and the state with an equal number of balls will be called the equilibrium state—even though it is not really an equilibrium in any dynamical sense. Let's start small, with just one ball—one flea jumping from one dog to another. Then we have a deterministic process: Rover, Queenie, Rover, Queenie, Rover, ... With two fleas, if we start with the state of both on Rover, then the probability of a transition to equilibrium is 1, whereas the probability of a transition from equilibrium to both on Rover is $\frac{1}{2}$. Suppose we have 1000 fleas and we are far from equilibrium—say 990 on Rover and 10 on Queenie. Then the probability of a transition toward equilibrium—an increase in Boltzmann entropy—is 99%, and the probability of a transition away from equilibrium—a decrease in Boltzmann entropy—is 1%. If one is *exactly* at equilibrium, however, a move away from equilibrium is certain. Baldly put, for Boltzmann entropy, the second law of thermodynamics—taken strictly—is false.

But the pet owners can hardly tell the difference between 500 fleas on each and 400 on one and 600 on the other. From the owners' point of view, the dog-flea system far from equilibrium tends towards equilibrium, and the system at equilibrium tends to stay there. It is also still true that if one starts at a state far from equilibrium and waits long enough, that state is certain to recur. The Poincare-Zermelo recurrence phenomenon obtains for the Ehrenfest urn model. But if there are a large number of fleas and the initial state is far from equilibrium, the mean time to recurrence is astronomical, and for a long time the process will look as if the second law of thermodynamics holds.[6]

There is another way to look at the model, which we describe here in subjective terms. Suppose the owners know that there are a lot of fleas and know about the transitions but do not know how the fleas were initially distributed. They now have to consider not a single dog-flea system evolving, but rather a set of systems.

Perhaps they think that, for all they know, initial states were equiprobable.[7] They know that the dogs had been together in the same kennel for a week. Then they might well expect with high probability that the dogs would be approximately at equilibrium, given that the fleas have had time to mix it up. (This can be calculated, and it doesn't take that long.[8])

Alternatively, suppose that they start with high initial probabilities of the systems being far from equilibrium and leave their dogs in the kennel for a day. They might well expect, with high probability, to find the dogs closer to equilibrium at the end of the day than at the beginning.

This is the point of view adopted by Gibbs and Maxwell. The focus is not on the time evolution of an individual system, but rather on the time evolution of a probability distribution over systems. Gibbs calls these probability distributions ensembles, as if we had many copies of an Ehrenfest urn model and were counting frequencies over this population. The populations are fictitious, and the ensembles are just a way of personifying probability distributions. An equilibrium is a stationary probability distribution.

In this way of looking at things, we can now say we approach equilibrium, and if we are there, we stay there. To contrast the two points of view, let us go back to the case of two fleas. There are three states: both on Rover, both on Queenie, or one on each. The process is a Markov chain with a unique limiting distribution, which is reached from any starting distribution. This places probability $\frac{1}{4}$ for both on Rover, probability $\frac{1}{4}$ for both on Queenie, and probability $\frac{1}{2}$ for one on each. This equilibrium, once reached, never changes. If we think in terms of ensembles, it is graphically clear. Consider four copies of the dogs, one copy with both fleas on Rover, one with both on Queenie, and two with one on each. Then the copy with both on Rover changes to one on each, as does that with both on Queenie. And one of the copies with one on each changes to both on Rover, while the other changes to both on Queenie. For every flea in one copy of the system that jumps one way, there is a flea in another copy that jumps the other. So when the probability distribution reaches equilibrium, it stays there, even though each individual system is fluctuating like mad.

The dramatic contrast between these two viewpoints is, of course, attenuated as the number of fleas is increased. Suppose we have 1000 fleas. Then in the Gibbs picture, the equilibrium distribution piles up close to half on each dog. Once at equilibrium, the ensemble does not change. In the Boltzmann picture, the system spends most of its time close to half on each dog, but fluctuates. And if one waits long enough, it will visit the state where all fleas are on one dog. Both points of view are correct; they are just looking at different things.

Everything here is clear, but in what sense is a gas like an Ehrenfest urn?

PROBABILITY, FREQUENCY, AND ERGODICITY

What is the probability at issue, and where does it come from? Boltzmann appears (for the most part) to be a frequentist. Probabilities of states are long run limiting frequencies. He believes that since the micro-dynamics is very fast, observables could be equated to long-run frequencies. But determining such long-run frequencies for an initial condition—or averages of such frequencies for a set of initial conditions—by analyzing the microdynamics is an impossible task. Boltzmann believed, however, that long-range time averages could be equated with averages over phase space, which are easy to compute. In fact he sometimes explains probability using both notions as if they were interchangeable.

In support of this assumption, he believes that the dynamics is such that, starting from every initial condition, the evolution of the system will carry it throughout every point in phase space.[9] This is his *ergodic hypothesis*. In later publications he acknowledged that speaking strictly, the ergodic hypothesis was false. He knew of counter-examples, such as particles bouncing back and forth between two parallel walls of the container, but considered them so unlikely as to be unimportant. So he really held what has been called the *quasi-ergodic hypothesis* that most, or almost all, initial conditions had the required character. But Boltzmann had no proof that the hypothesis held for the model of gas at issue. And, furthermore, he lacked a proof from the ergodic hypothesis to some sort of statistical *H*-theorem. The Ehrenfests had shown how a statistical version of the second law that Boltzmann had envisioned escaped the objections of Loschmidt, Poincaré, and Zermelo—but it lacked a mathematical foundation.

VON NEUMANN AND BIRKHOFF ON ERGODICITY

The proper notion of ergodicity for equating time averages and phase-space averages was formulated by John von Neumann and George Birkhoff around 1930. The correct mathematical setting is a probability space, as in Kolmogorov (chapter 5), equipped with a dynamics. For discrete time, the dynamics can be thought of as a transformation on the space that respects the probability space in that it carries measurable sets to measurable sets.[10]

If the transformation preserves the probability—$P(S) = PT^{-1}(S)$ for every measurable set S—then the measure is *invariant* with respect to the dynamics. A *measurable set is invariant* if it maps onto itself except for a set of measure 0.[11] An invariant measure is *ergodic* if the only invariant sets are those of measure 1 and measure 0. Or, if the measure is taken as given, the dynamical system itself is said to be ergodic with respect to the measure.

Birkhoff proved that if a dynamical system is ergodic, then, except for a set of points of probability 0, the average of a measurable function over the probability space equals the limiting time average.

$$\int f \, dp = \lim_{n \to \infty} \frac{1}{n} \{ f(T(x)) + \cdots + f(T_n(x)) \}.$$

For Boltzmann's model of a gas the uniform measure[12] is invariant with respect to the dynamics.[13] This leaves us with these questions: (1) Is the system ergodic? and (2) Does ergodicity explain the second law?

Even at the present, the first question is still unresolved. In 1963 Sinai[14] announced a proof that Boltzmann's model of n hard spheres was ergodic, but the general claim was retracted in 1987.[15] There is a proof for three hard spheres, and there are other partial results. Boltzmann's model may well be ergodic, but, as we write, this is not yet proven.

Suppose ergodicity can be proved. Would that, in itself, be enough to provide the desired explanations of thermodynamic phenomena? How much does equality of time and phase averages give us? Consider two well-known examples of ergodic systems.

Example 1: The first has only two states, call them H and T. The measurable sets consist of the unit sets of the states, the universal set and the null set. The probability measure gives each probability $\frac{1}{2}$. The dynamics deterministically maps H to T and T to H:

$$HTHTHTHTHTHTHTHT \ldots .$$

The measure is invariant. And the only invariant sets are the universal sets and the null set, so the system is ergodic. It is true that *if all we could observe were limiting relative frequencies*, the system would appear to forget its initial state and look just like coin tossing. But since we can see more, it forgets nothing and is perfectly predictable.

Example 2: The state space is the half-open interval $[0, 1)$ with Borel sets* as measurable sets. Probability is uniform (Lebesgue measure).

---

*Remember from chapter 5, these are the sets gotten by starting with intervals and closing under countable Boolean operations.

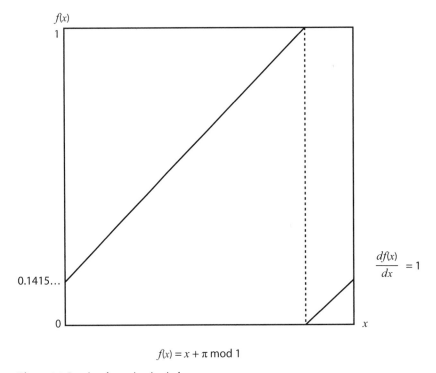

$$f(x) = x + \pi \bmod 1$$

Figure 9.1. Irrational rotation in circle

The dynamic maps x to $x + \pi$ modulo 1, as shown in figure 9.1. Thus the dynamic is an irrational rotation of the circle. Since the dynamic is a translation, the measure is invariant. The only invariant sets are the null set and the universal set.

POINCARÉ

In the foregoing examples, ergodicity seems a rather weak requirement. The key idea for strengthening it comes from Poincaré. He puts it this way in a popular essay, *Science and Method* (Tr. Francis Maitland (London: Thomas Nelson, 1914): Part 1, Chapter 4, p. 68):

> If we knew exactly the laws of nature and the situation of the universe at the initial moment, we could predict exactly the situation of that same universe at a succeeding moment. But even if

it were the case that the natural laws had no longer any secret for us, we could still only know the initial situation *approximately*. If that enabled us to predict the succeeding situation with *the same approximation*, that is all we require, and we should say that the phenomenon had been predicted, that it is governed by laws. But it is not always so; it may happen that small differences in the initial conditions produce very great ones in the final phenomena. A small error in the former will produce an enormous error in the latter. Prediction becomes impossible, and we have the fortuitous phenomenon.

Poincaré is describing the sensitive dependence on initial conditions that he found in his chaotic dynamics in the three-body problem. It is evident from just his description that our example of the irrational rotation of the circle does not exhibit this sensitive dependence. Points that start close remain close.

The way to quantify sensitivity to initial conditions was introduced in 1892 by Aleksandr Lyapunov in his doctoral dissertation, *The General Problem of the Stability of Motion*.[16] Lyapunov exponents measure the average exponential divergence of near points.

In our simple case of maps of the unit interval onto itself $x_{n+1} = f(x_n)$, the rate of separation of nearby points on one iteration is given by the log of the derivative $f'(x)$. Averaging over iterations gives us the Lyapunov exponent for the orbit as

$$\lim_{N \to \infty} \frac{1}{N} \sum_{n=1 \text{ to } n=N} \log |f'(x_n)|.$$

We get a Lyapunov exponent for the whole system by taking an expectation.

In example 2, given earlier, an irrational rotation of the circle, $f'(x)$, is the same everywhere and equal to 1. Thus the Lyapunov exponent is $\log(1) = 0$, indicating that there is no exponential divergence. We have ergodicity without sensitive dependence on initial conditions.

For an interesting example where we *do* have sensitive dependence, consider the Bernoulli shift, shown in figure 9.2:

$$f(x) = 2x \bmod 1.$$

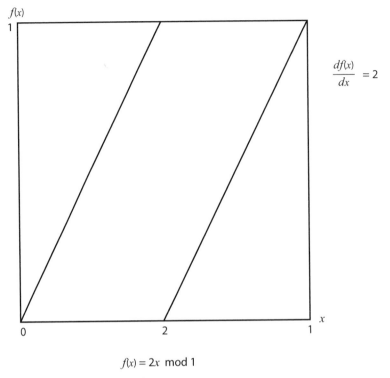

$$f(x) = 2x \mod 1$$

Figure 9.2. Map for Bernoulli shift on $[0, 1)$

That is, for x in $[0, 1)$, if $x < \frac{1}{2}, f(x) = 2x$; otherwise $f(x) = 2x - 1$. Taking our probability measure as the uniform measure on $[0,1)$, you can see, as shown in figure 9.3, that the measure of an interval is equal to the measure of its inverse image. The shift is measure preserving.

Recall (from Borel in chapter 5) how infinite sequences of heads and tails (0s and 1s) can be considered as the binary expressions of real numbers in the unit interval, $[0,1)$. Viewed from this perspective, $f(x)$ is the point gotten by taking the binary expression of x and throwing away the first digit, thus the name *Bernoulli shift*. The representation tells us that the only fixed point (in binary) is $0.000000000000000\ldots$ and that cycles correspond to other rational numbers. There are lots of invariant sets, for example, $\{\frac{1}{3}, \frac{2}{3}\}$, but they all have measure 0 or measure 1. The system is ergodic.

For almost all $x, f'(x)$ exists and is equal to 2. An initial ε separation of two points becomes $2\varepsilon, 4\varepsilon, 8\varepsilon, \ldots$ as the map is iterated. The

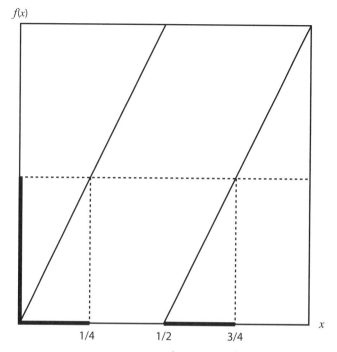

Figure 9.3. Bernoulli shift. Inverse image of $[0, \frac{1}{2})$ as measure $\frac{1}{2}$.

Lyapunov exponent is $\log(2)$. This is clearly an example of sensitive dependence on initial conditions. It is apparent that this system forgets its initial state in a sense much stronger than that guaranteed by mere ergodicity. The probability that the system is in $[0, 0.5)$—heads on the first toss—is $\frac{1}{2}$. The probability that the system is in $[0, 0.5)$ after one iteration of the map—heads on the second toss—given that it started in $[0, 0.5)$ is again $\frac{1}{2}$. *If all we could see in the Bernoulli shift dynamics was whether the system was in the left or right half of the unit interval, it would look just like tossing a fair coin.* In this way the Bernoulli shift models Bernoulli trials.

If a dynamical system has sensitive dependence on initial conditions, then, under mild conditions, a prior probability on the initial conditions washes out, leading to the Boltzman-Liouville uniform distribution.[17]

THE ERGODIC HIERARCHY

There are intermediate cases between the immediate forgetting of the Bernoulli shift and the complete lack of forgetting that is possible with mere ergodicity. These are ordered in an *ergodic hierarchy*. There is the concept of a *mixing* dynamical system. Consider two measurable sets A and B. Initially they may not be independent (the measure of their intersection may be different from the product of their measures). Then the proposition that a point is in A carries information about whether it is in B. Now let the points in the intersection of A and B evolve according to the dynamics, for n steps. This gives a new set of points with a new measure. If this measure is just the product of the initial measure of A and that of B, the system has "forgotten" the initial correlation. The definition of a *mixing* dynamical system is that it forgets correlations in the limit.[18]

Mixing systems forget correlations, but it may take forever. There is another step in the ergodic hierarchy, that of a K-system (or Kolmogorov system). K-systems fall between Bernoulli systems, and mixing systems. So the main levels of the hierarchy are

Bernoulli,
K-system,
mixing,
ergodic,

with higher-level properties implying lower-level ones. Poincaré's sensitive dependence on initial conditions tells us something about the *rate* at which the system forgets the past. In chaotic dynamical systems—those with a positive Lyapunov exponent—close points diverge exponentially fast. Such chaotic systems have positive Kolmogorov-Sinai entropy and must be at least K-systems in the hierarchy.[19]

BOLTZMANN REDUX

It is evident that Boltzmann's program requires something stronger for a Boltzmann gas than mere ergodicity. In fact, what Sinai proved in 1970 was that for two spheres (in two dimensions, i.e., two disks) on a torus, the model is not just ergodic, but it is also a K-system. Subsequent research aims at showing that a Boltzmann model gas is also a K-system. This would indeed provide a mathematical foundation for Boltzmann's program. One of the leading researchers, in a symposium on the 150th anniversary of Boltzmann's birthday (1994) wrote: "After more than one hundred years, ergodicity is still not established in the simplest mechanical model, in the system of elastic hard balls . . . ."[20] As we write, this is still true.

The problem remains as a mathematical problem, but it has receded from a central place in statistical mechanics. Simulations suggest a Boltzmann gas is ergodic and rapidly mixing, even if it hasn't been proved. One can just postulate the microcanonical distribution that Boltzmann sought to justify, use it, and get results that match some empirical phenomona. And, after all, real gases aren't Boltzmann gases.

QUANTUM MECHANICS

To even its most practiced professionals, quantum mechanics is mysterious. Consider one of the simplest problems; a quantum particle evolving on a one-dimensional line. The position of the particle at time t is described by a *wave function*, $\Psi(x, t)$. Born's statistical interpretation of the wave function says that $|\Psi(x, t)|^2$ is the probability of finding the particle at position x at time t. (More precisely, it is the probability density at x at time t.)

What kind of probability can this be? One standard textbook presentation favors a frequency interpretation: if the experiment is repeated infinitely often under identical circumstances, $|\Psi(x, t)|^2$ integrated between a and b gives the limiting proportion of times that the actual observation is between a and b. This fiction is even more jarring in the quantum setting than in the classical setting. How are these infinitely many trials supposed to be coupled? (Do their particles interact?) All the weaknesses of the classical frequentist apply here.

To be clear, we have no complaints about the standard formalism of quantum mechanics. It explains a dazzling array of observable features of the real world with astonishing accuracy. It also explains a host of very counterintuitive phenomena special to microscopic observables—things like the Heisenberg uncertainty principle (position and velocity cannot be simultaneously observed to arbitrary accuracy).

Does quantum mechanics need a different notion of probability? We think not. As long as we confine ourselves to the level of experiment and result and eschew extra metaphysics the issues are basically the same as in the classical setting. Long-run relative frequencies require a somewhat strained hypothetical counterfactual interpretation, which leaves them not being real frequencies at all. We can posit physical chances computable from the quantum state, which are primitives feeding into Bayes' theorem. We can consider robust degrees of belief based on all the evidence. We find the last point of view as congenial in the quantum setting as in the classical one. The difference with classical games of chance is that in those cases we now have a physics that tells us how a microspecification of the system leads to determinism. In quantum theory, the physicist is confined to a position quite analogous to that of the naive, pre-Newton, dice player.[21, 22, 23]

The formalism of quantum mechanics led immediately to new mysteries in physics.

NONLOCALITY

According to the orthodox view of quantum mechanics, the theory is fundamentally probabilistic. Quantum mechanics "sets the world at chance." But quantum probabilities have mysterious aspects. They can embody something that looks, on the face of it, as a spooky (Einstein's word is "spukhafte") action at a distance. And this apparent action can even be deterministic.

The fundamental issue was raised in a thought experiment by Einstein, Podolsky, and Rosen in 1935.[24] Two particles can be prepared in a quantum state (an entangled state) and then separated in space, such that measurement on one affects the probabilities of measurement results on its distant partner. This is the nonlocality that struck Einstein as spooky. Einstein, Podolsky, and Rosen preferred the alternative that there is some deeper, hidden level of reality that does act locally, and that explains the quantum phenomena.

John Stewart Bell

The EPR argument was sharpened by John Bell in 1964.[25] Two electrons can be created in an entangled state[26] and then widely separated such that according to quantum theory the following is true. If the quantum spin of both electrons is measured along the same axis (by passing them through an appropriate magnetic field), then the results are perfectly anticorrelated. If one gets spin-up on the left electron, one gets spin-down on the right. If one gets spin-down on the left electron, one gets spin-up on the right. These two possibilities each have probability $\frac{1}{2}$. How does the right electron "know" how the measurement on the left electron came out?

Locality is restored if each electron carries hidden information, imparted at the source, about whether to go up or down. The source creates electrons in two types of pairs with equal probability. One type has information that tells the left electron to go up and the right to go down; the other has just the opposite. Now spooky nonlocality has been eliminated. But note that to do so, the hidden property must act deterministically in affecting the measurement result.

Now spin may be measured on any axis around the clock, 12–6, 9–3,—any axis. So each electron must carry deterministic information about how to act under any such measurement, and the source must produce electrons with different hidden properties in such a way as to preserve the anticorrelation on any axis and to keep the probabilities of up or down equal. This is mathematically possible.

But quantum theory also specifies the probabilities of measurement results when the spin measurements are not on the same axis. For our local hidden information model to agree with quantum theory, the source must produce pairs with frequencies that match the quantum mechanical predictions for all such measurements. It turns out, as Bell showed, that this is mathematically impossible.

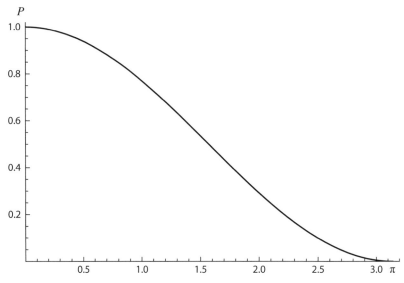

Figure 9.4. Probability of opposite results as a function of angle θ

Suppose that the axis of detector on the right is set at an angle, θ, to the oriented axis of the detector on the left. Then quantum mechanics gives the probability of finding opposite results, up-down or down-up as

$$(1 + \cos \theta)/2,$$

as shown in figure 9.4. If θ is zero, the probability is 1, the deterministic case just discussed. If θ is $\pi/2$, so that the axes are at right angles, the probability is $\frac{1}{2}$. Left and right are uncorrelated. If θ is π, the probability is 0. Again we have perfect correlation, but the orientation of the axes has changed, so that what counts as "up" on one side now counts as "down." If the angle between the oriented axes is $2\pi/3$, one-third of the way around the circle, then the probability of anticorrelation is $\frac{1}{4}$. This is the case that we use to get a maximal inconsistency between the envisioned hidden causes account and quantum theory.

Consider three axes, A, B, C, equally spaced such that the angle between any two is $2\pi/3$. Measurements are made at the distant locations left and right on any combination of left and right axes. A measurement yields either Up or Down as a result. Remember that our locality hypothesis was that the source emits electrons that carry local information telling them how to behave, Up or Down, in every measurement situation, and emits them in proportions that give the quantum mechanical statistics.

Consider the measurements at A, B, and C on the left. There are only two possible outcomes, so the measurement results must match in at least two cases. Thus, just by logic, we have

$$P(M_A = M_B) + P(M_B = M_C) + P(M_C = M_A) \geq 1.$$

Now consider the deterministic relation of measurements on the left and right that holds when they are on the same axis. If M_B on the left is Up, then M_B on the right is Down, and conversely. So $M_{A \text{ on Left}} = M_{B \text{ on Left}}$ is equivalent to $M_{A \text{ on Left}} \neq M_{B \text{ on Right}}$. Likewise for the pairs of axes $\langle B, C \rangle$ and $\langle C, A \rangle$. But quantum theory told us that

$$P(M_{A \text{ on left}} \neq M_{B \text{ on right}}) = \tfrac{1}{4}.$$

And, likewise,

$$P(M_{B \text{ on left}} \neq M_{C \text{ on right}}) = \tfrac{1}{4}$$

and

$$P(M_{C \text{ on left}} \neq M_{A \text{ on right}}) = \tfrac{1}{4}.$$

Thus, our hidden information theory can pull off the trick of preserving locality only if

$$\tfrac{1}{4} + \tfrac{1}{4} + \tfrac{1}{4} \geq 1.$$

It just can't be done.

That was a nice clean thought experiment. What about real experiments? The best experiments use photons rather than electrons.[27] Some make the measurements when the photons are more than 10 kilometers apart.[28] Some are able to choose the orientation of the detectors while the photons are in transit.[29] More and more stringent experiments continue to be done. These experiments have all come out in favor of quantum theory.[30] We are left with the quantum probabilities.[31]

QUANTUM PROBABILITIES REDUX

With the EPR experiment under our belt, we can reiterate what we claimed before. Nothing in our treatment of EPR moves us outside classical probability theory. Carrying out a physical measurement results in the information that a certain result was gotten by carrying out a certain measurement procedure. This information is incorporated into our degrees of belief by Bayesian conditioning, like any other new information.

Suppose the measurement axes A, B, and C are chosen independently with equal probability by some local quantum randomizer, as indeed they are in one of the rigorous experimental tests of Bell. Then we have a perfectly classical probability space whose points are specifications of

<measurement orientation on left, measurement orientation on right,

measurement result on left, measurement result right>.

Combinations of results conditional on measurement settings are given by quantum theory. The nine possible measurement settings are equiprobable. Probabilities of the points are given by the theorem on total probability. What caused the trouble was not classical probability, but the assumption of locality. The second thing to note is that the quantum nonlocality that was demonstrated does not allow probabilistic signaling between distant experimenters. The experimenter on the left can choose the orientation of the measurement device, but she cannot choose the measurement results. Whatever the orientation on the left, the probability of getting Up on the right equals the probability of getting Down.

QUANTUM CHAOS

Chaos justified the use of probability in a classical setting. In particular, a chaotic dynamic was hypothesized (but not quite yet proved) for a Boltzmann gas. A real gas—real everything—is presumably quantum mechanical. We should reconsider our classical picture in the

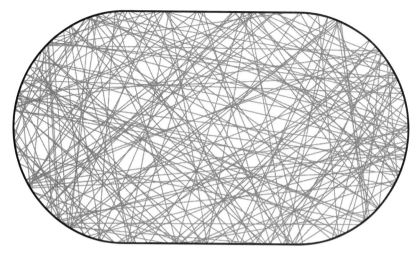

Figure 9.5. The stadium exhibits chaos

light of quantum mechanics. It is, therefore, somewhat unsettling that in quantum mechanics chaotic dynamics are impossible!* But quantum mechanics is supposed to behave like Newtonian mechanics in the macroscopic domain. Also, classical Newtonian dynamics allows chaos and predicts it in many situations. How are we to resolve these apparent contradictions?[32]

Consider a Boltzmann gas with just one molecule, a classical billiard. The billiard bounces off the wall of an enclosure, with the angle of incidence equal to the angle of reflection. Is the dynamical system ergodic? If so, where is it in the ergodic hierarchy? It depends on the shape of the enclosure. If it is circular, the system is not ergodic. If it is a stadium—two half-circles joined by two parallel straight lines, as shown in figure 9.5—it has been proved to be ergodic.[33] Almost every orbit fills the whole space. (There exist exceptional "bouncing ball" orbits, where it just bounces between the parallel sides, meeting the sides at right angles, but these have measure zero.) If the enclosure is heart shaped—a cardioid, as shown in figure 9.6—then it is as chaotic as can be. Any two nearby orbits separate exponentially fast. Every orbit fills the whole space. Now suppose our billiard is an electron, delicately confined within a cavity so as to insulate it from the environment and to approximate the

---

*Because the dynamics are linear.

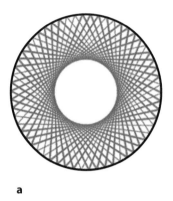
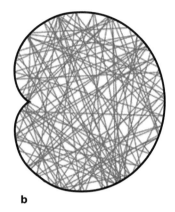

a   b

Figure 9.6. (a) is not chaotic; (b) is chaotic

foregoing. We will not see the same picture. We cannot, of course, look at orbits of a billiard in physical space. They do not exist.

We can look at (computationally) the steady states that the system can assume at various energy levels. For a classically chaotic system we might hope to find probability of being at a physical point, if we look, being uniformly distributed (and likewise for momentum). At low energy levels, this is not at all the case. But in the limit, as energy approaches infinity, it is. It is a theorem:

Classical ergodicity implies quantum ergodicity.[34, 35]

To be sure, we are not really at this (semiclassical) limit, even at the macroscopic scale. It seems that chaotic dynamics has often been observed involving large, massive objects. Then shouldn't, in theory, quantum periodicity assert itself? It should, over long-enough periods of time, if the system remained isolated. But exquisite isolation over extended periods of time is not easy to come by on the macroscopic level. And interactions with the environment destroy the coherence of the quantum subsystem that would produce the periodicity. This is known as decoherence.[36]

Quantum mechanics does not nullify the importance of classical chaotic dynamics for reasoning about physical chance. The picture becomes more nuanced as we zoom in. The theory becomes more complex. The reasons to think probabilistically about the physical world are multiplied.

SUMMING UP

With the statistical mechanical explanation of thermodynamics, probability acquired a new importance in physics. This led to difficulties, puzzles, and paradoxes that took some time to resolve but which ultimately led to a conceptual clarification of the theory. The path led through the development of ergodic theory, the development of Poincaré's idea of sensitive dependence on initial conditions into chaos theory, and incorporation of both strands of thought into the ergodic hierarchy. The background physics is deterministic, but a little uncertainty about initial conditions leads to large uncertainty about predictions.

With quantum theory, the fundamental physics is not deterministic, and prediction appears to be uncertain at a deeper level. New counterintuitive correlations of measurement outcomes at a distance appear. In some attempts to restore some intuitions, deterministic hidden-variable theories have been developed, but they produce no new predictions and retain the counterintuitive character of nonlocality. Quantum chaos is, in a strict sense, impossible, but quantum systems that are not perfectly isolated can exhibit behavior that looks like classical chaos.

Many of the mathematical questions raised by the physics remain to be settled, from quantum chaos to the simplest toy model of a Boltzmann gas. Nevertheless, from classical to quantum physics, we are presented with a world of chance. We believe that throughout classical and quantum physics, the same philosophical problems regarding the nature of chance arise, and the same considerations hold good.

APPENDIX. QUANTUM METAPHYSICS:
A PEEK INTO PANDORA'S BOX

In our discussion of Einstein-Podolsky-Rosen and Bell's theorem, we stayed at a nice safe operational level. Many have found the operational level unsatisfactory and have tried to formulate a more satisfying metaphysical theory that nevertheless delivers the same quantum-mechanical results at the operational level. The literature is vast and complex, and we have no intention of attempting a full discussion of it here. This appendix is just meant to point to some of the issues. We call it metaphysics, not to be derisive, but because the different positions do not, and are not meant to, generate any new physics.

On the operational level we have experiments and results, measurements performed, and outcomes. Probabilities of outcomes conditional on measurement performed are given by the quantum mechanical formalism. Measurements are performed and outcomes are observed. Quantum mechanical probabilities are validated by the kind of reasoning introduced by Thomas Bayes. Performing a measurement has physical consequences other than delivering measurement results. The formalism accounts for these by the collapse of the wave packet. What bothers people, from Einstein to the present, is that there seems to be no satisfying account of the measurement process. Collapse of the wave packet is a strange physical process, discontinuously interrupting the smooth evolution of the wave function according to Schrödinger's equation. What qualifies something to be a measurement? What's going on behind the scenes?[37]

Here are two main approaches. (There are many others.) The first is the de Broglie-Bohm[38] pilot wave theory. All measurements are position measurements. (Think of pointer readings for all other measurements.) Everything always has a position, even when it is not being measured. What happens when you are not looking arranges this to give the quantum mechanical probabilities. This is effected by a wave, which evolves according to the Schrödinger equation. The wave guides the particles so as to deliver the desired result. The two components of the theory, the quantum wave and the deterministic particle, have to cooperate in order to reproduce the quantum mechanical predictions. The character of the quantum mechanical probabilities in this picture is not much different from the character of probabilities of the outcomes of rolls of dice for classical gambler.

What happens to the guiding wave during a measurement? One must say something that explains wiping out interference. Basically, the story is that wave function does not collapse; it just looks like it does.[39]

The second approach—due to Everett[40]—seems metaphysically quite different. Instead of the dualism of the Bohmian picture, we have here only the wave function. There is a wave function for the whole universe, and it evolves according to the usual dynamics. It does not collapse. It is all that there is.

What is this account to make of measurement?[41] What is being measured? What about the supposed physical effects of the wave function collapse? The answers in the Everett picture should have a familiar ring. The wave function does not collapse; it only looks like it does. As in Bohmian measurement, a key element in the account of why it looks like the wave function collapses is *decoherence*—caused by the system not remaining isolated from the environment.

The system consisting of the observer and the observed becomes entangled in such a way that relative to the state of the observer, the observed has assumed a definite state. For instance, assume a *perfect* spin measurement. After the measurement event, the observer-observed system is in a superposition of "spin-up, observer records spin-up" and "spin-down, observer records spin-down." This explains the metaphor of splitting into two possible worlds and the appellation many-worlds interpretation of quantum mechanics. (But if no measurement is perfect, the "splitting" is never perfectly clean.)

What is the probability given by the Born rules the probability *of*? The account of measurement in Everett is an explanation of the *appearance* of a collapse of the wave function. The probability is then the probability of it *appearing* to collapse one way or another. How this interpretation of the Born rule is justified is a matter of some controversy.[42] These are issues that the reader may wish to explore. In the body of the chapter, we confined ourselves to the probabilities at the core operational level of quantum mechanics, which each approach aims to recapture.

David Hume

CHAPTER 10

INDUCTION

Will the future be like the past? Not always. "The man who has fed the chicken every day throughout its life at last wrings its neck instead . . . ." Too bad for the chicken, but it might prompt philosophical reflection in those so inclined—as Bertrand Russell intended.

> The mere fact that something has happened a certain number of times causes animals and men to expect that it will happen again. Thus our instincts certainly cause us to believe the sun will rise to-morrow, but we may be in no better a position than the chicken which unexpectedly has its neck wrung. We have therefore to distinguish the fact that past uniformities *cause* expectations as to the future, from the question whether there is any reasonable ground for giving weight to such expectations after the question of their validity has been raised. (Russell, *The Problems of Philosophy*, chapter VI)[1]

The question is not so easy to answer as one might, at first thought, believe. If you already have the answer, we beg you to, nevertheless, persist a bit and read on.

This problem of induction was first raised in full force by the philosopher David Hume. Hume's great idea is that there is a problem of understanding and validating inductive reasoning. Why should you take this seriously as a great idea?

In 1967 L. J. Savage, the great Bayesian author of *The Foundations of Statistics*, wrote

> Some of the most untrustworthy of philosophical demonstrations have proved the most valuable. Warriors can overtake tortoises; yet Zeno convinces us that there is more to motion than meets the eye. And the importance of Hume's argument against induction—the keynote of this symposium—is undoubted; though perhaps most philosophers view it, like the arguments of Zeno, only as a challenge to search out a manifest fallacy. Some of us, however, find Hume's conclusion not paradoxical but close to the mark.[2]

De Finetti was inspired by Hume, and he also thinks that Hume was fundamentally right but also that he has resolved Hume's problem: "I think that the way in which I have sought to resolve the problem of inductive reasoning is nothing but a translation into logico-mathematical terms of Hume's ideas. . . ."[3] We saw that Bayes and Price wanted to answer Hume. Karl Popper, a prominent philosopher in the twentieth century, thought that Hume was unanswerable and that, as a consequence, inductive inference was impossible. In retrospect, most of what preceded this chapter can be seen as a story of brilliant attempts to come to terms with the problem of induction raised by Hume.

Let us keep an open mind and review in what way, if any, probability theory speaks to inductive skepticism.

HUME

The classic modern statement of inductive skepticism comes from David Hume, although he reminds us of its ancient sources. How, asks

Hume, could we justify inductive reasoning? Inductive reasoning is not justified by relations of ideas—by mathematical deduction:

> That the sun will *not* rise tomorrow is no less intelligible a proposition than that it will rise. We should in vain, therefore, attempt to demonstrate its falsehood.[4]

Think of the world as like a movie. You can take one movie, cut it, and splice it to any entirely different one. We could be at the splice. The scenario is not inconsistent. Pure mathematics cannot justify induction. But to try to justify inductive reasoning by inductive reasoning is to beg the question:

> It is impossible, therefore, that any arguments from experience can prove this resemblance of the past to the future, since all these arguments are founded on the supposition of that resemblance.[5]

Then, by Hume's lights, there is no avenue left. This remarkably simple argument has occupied philosophers from his time to the present.

KANT

Kant took Hume seriously and set out to solve the problem of induction. Some philosophers believe that he succeeded. He says that he has. But it is not so easy, even for Kant specialists, to say just how Kant's solution goes. If you just read Hume, it is easy to see what he is saying. With Kant—not so much.[6] Whole careers are spent trying to puzzle out what his position really is.

Hume emphasizes that our innate psychology disposes us to causal and inductive formation of expectations (like Russell's chicken). With Kant, this Humean psychology first becomes codified in a priori rules and then somehow transmuted into knowledge of a new kind—the *synthetic* a priori. Sorry, but we just do not see it. With apologies to Kantians, we agree with the philosopher C. D. Broad:

There is a skeleton in the cupboard of Inductive Logic, which Bacon never suspected and Hume first exposed to view. Kant conducted the most elaborate funeral in history, and called Heaven and Earth and the Noumena under the Earth to witness that the skeleton was finally disposed of. But, when the dust of the funeral procession had subsided and the last strains of the Transcendental Organ had died away, the coffin was found to be empty and the skeleton in its old place.[7]

POPPER

One might simply read Hume and give up. This, in fact, is the position taken by Sir Karl Popper. Popper was a Viennese philosopher who moved to England and spent many years at the London School of Economics. In a very influential book on the logic of science,[8] he wrote

> Yet if we want to find a way of justifying inductive inferences, we must first try to establish a *principle of induction*. . . .
>
>   Now this principle of induction cannot be a purely logical truth . . .
>
>   . . . if we try to regard its truth as known from experience, then the very same problems which occasioned its introduction will arise all over again.

So far, Popper has simply rephrased Hume's argument. But then he immediately concludes that Hume has shown the impossibility of inductive logic:

> My own view is that the various difficulties of inductive logic here sketched are insurmountable.[9]

Popper goes on to develop an account of science in which there are no inductive inferences; it is supposed that all inference is purely deductive. Experimental predictions are deduced from theories, if observation proves them wrong, the theory is refuted and that is all there is to it.

Karl Popper

This view has a certain superficial plausibility, and some scientists pay lip service to it. But if one tries to actually carry out the deduction of an experimental prediction, one will find that these deductions are deductions in the sense of Sherlock Holmes—that is to say that one finds chains of plausible inductive inferences. As an exercise, one might try to deduce observation from theory in one of the physical theories that we discussed in chapter 9. Putting many other difficulties to the side, we note that at best we deduce something about chances from the theories. And, at best, we observe finite frequencies. If we are careful, and don't fall prey to Bernoulli's swindle, we must find inductive inference.

GRADES OF INDUCTIVE SKEPTICISM

One might well ask why Popper did not think to apply the tropes of Agrippa, the ancient skeptic,[10] to deductive reasoning, just as Hume had applied them to inductive reasoning.* There is an infinite regress—mathematics is justified by set theory, whose consistency is

---

*Agrippa applied them to all kinds of reasoning. Thus, in the account of Diogenes: "And in order that we may know that an argument constitutes a demonstration, we require a criterion; but again, in order that we may know that it is a criterion we require a demonstration . . . ."

proved by stronger set theory, and so on—or, regarding mathematics as a whole, a circularity.

Or why not ask why one should accept any argument at all? Is giving an argument why one should do so, begging the question? *Trying to answer a thoroughgoing skeptic is a fool's game.* You might well ask if we are engaged in such a game in this chapter.

But it is possible, and sometimes quite reasonable, to be skeptical about some things but not others. Thus, there are *grades of inductive skepticism*, which differ in what the skeptic calls into question and what he is willing to accept.[11] For each grade, a discussion of whether such a skeptic's doubts are justified *in his own terms* might actually be worthwhile. Put in that way, we can see—as we said—that we have already been engaged in a discussion of inductive skepticism throughout this book. This discussion began almost contemporaneously with the advent of serious probability theory. Here we review what we have learned from this perspective.

BAYES-LAPLACE

In chapter 2 we saw that Jacob Bernoulli thought that he had solved the problem of inferring chances from data with his law of large numbers:

> [W]hat you cannot deduce a priori, you can at least deduce a posteriori—i.e. you will be able to make a deduction from the many observed outcomes of similar events. For it may be presumed that every single thing is able to happen and not to happen in as many cases as it was previously observed to have happened or not to have happened in like circumstances.[12, 13]

Bernoulli had proved that with enough trials it would become "morally certain" that the frequency would be approximately equal to the true chance. If x is approximately equal to y, then y is approximately equal to x. So, after a large number of trials, we can take the true chances to be approximately equal to the observed frequencies.

This informal argument gains an air of plausibility by concealing difficulties behind a cloak of moral certainty and approximate equality. This is what we called *Bernoulli's swindle* in chapter 4. At this point, what is called for is a healthy skepticism, which would clear the way for a real analysis of inductive inference.

Thomas Bayes and Richard Price saw that Bernoulli's argument did not resolve the problem. Price gives a precise diagnosis in his preface to Bayes' essay:

> Mr. De Moivre . . . has, after Bernoulli, and to a greater degree of exactness, given rules to find the probability there is, that if a very great number of trials be made concerning any event, the proportion of the number of times it will happen to the number of times it will fail, in those trials, should differ less by small assigned limits from the proportion of the probability of its happening to the probability of its failing in one single trial.
>
> But I know of no person who has shown how to deduce the solution to the converse problem to this; namely, "the number of times an unknown event has happened and failed being given, to find the chance that the probability of its happening should lie somewhere between two named degrees of probability.[14]

In the section of the *Enquiry* devoted to probability, David Hume wrote:

> But where different effects have been found to follow from causes, which are to *appearance* exactly similar, all these various effects must occur to the mind in transferring the past to the future, and enter into our consideration, when we determine the probability of the event. Though we give the preference to that which has been found most usual, and believe that this effect will exist, we must not overlook the other effects, but must assign to each of them a particular weight and authority, in proportion as we have found it to be more or less frequent. . . .
>
> Let any one try to account for this operation of the mind upon any of the received systems of philosophy, and he will be sensible of the difficulty.[15]

Although they thought otherwise, neither Bernoulli nor de Moivre had given an answer to Hume.

An answer was given by Bayes and generalized by Laplace—the story that we told in chapter 6. Bayes was—to the best of our knowledge—addressing both Bernoulli's swindle and Hume's general skepticism. That was the view of Price, who was a friend of both Bayes and Hume.[16] We recall from chapter 6 the title page affixed by Price to reprints of Bayes' essay: *A Method for Calculating the Exact Probability of All Conclusions based on Induction*.[17]

Bayes had not given the predictive probabilities—the probabilities for outcomes in the next trial—for which Hume had asked an explanation. This step was taken by Laplace in his remarkable essay of 1774[18] (published at the age of 25). Assuming the uniform prior, he proves his famous rule of succession. Given p successes in $p+q$ trials, the probability of a success in the next trial is

$$\frac{p+1}{p+q+2}.$$

More generally, he considered the predictive distribution for m successes in $m+n$ additional trials given the evidence of p successes in $p+q$ trials. He shows that if the number of data $(p+q)$ is large and the number of future trials at issue $(m+n)$ is small, the result is close to taking the observed frequency as giving the chances, as conjectured by Hume (D. Hume, *An Enquiry Concerning Human Understanding* (London: Millar, 1748/1777). Section VI, "Of Probability"): "*. . . assign to each of them a particular weight and authority, in proportion as we have found it to be more or less frequent. . . .*"

But he also feels bound to point out that this is not the case if the number of trials predicted is also large: "and it seems to me essential to note this." We cannot resist a slight digression to compare this with the position taken by the philosopher Hans Reichenbach in the twentieth century. Reichenbach was a leading philosopher of science. He was a founder of the Berlin Circle, a group dedicated to promoting scientific philosophy. Von Mises was a member.[19] He was a rival of Popper and insisted that there was more to science than deductive reasoning. His approach to induction is still thought of as a serious contender by many philosophers. Reichenbach, with many, emigrated to escape the Nazis. He moved from Berlin to Istanbul (with von Mises) and then to UCLA, where he remained until his death.

Reichenbach, like von Mises, was a frequentist, but their theories differ. Von Mises does not advance any theory of induction. But

Reichenbach defends a rule that guesses, or "posits," the sample frequencies to be the limiting frequencies. Posits are *treated as true* until revised. Treating them as true means treating them as if they were a known chance:

> We posit h^n (i.e., the sample frequency) as the value of the limit, i.e., we wager on h^n just as we wager on the side of a die.[20]

If the number of data, $p+q$, is small and the number of future trials $m+n$ is large, Reichenbach would be led to bizarre betting behavior. To make the point in the extreme, consider bets on limiting relative frequencies. After 6 heads in 10 tosses, Reichenbach would be willing to bet his fortune against a penny that the limiting relative frequency would be 0.6. After one more trial he would be just sure of another limiting relative frequency. The same sort of thing happens, only to a lesser extent, if betting is on a large number of future trials. We believe that Reichenbach should have paid more attention to Laplace.

Laplace has more to offer in this essay. He showed what is now called *Bayesian consistency*:[21]

> One can suppose that the numbers p and q are so large that it becomes as close to certainty as one wishes that the ratio of the number of white tickets to the total number of tickets contained in the urn is included between the two limits $p/(p+q-w)$ and $p/(p+q+w)$. And w can be supposed less than any given quantity.[22]

Bayes-Laplace inference converges to the true chances.

Given their assumptions, Bayes and Laplace show that Bernoulli's conclusion was right. We can infer the approximate chances a posteriori. And, in this setting, *they do give an answer to Hume.* They show when, and in what sense, it is rational to believe that the future is like the past. They explain the proper use of past frequencies in calculating "weights" for future events.

But there are assumptions. We can be a little more skeptical and question these assumptions.

WHAT ABOUT THE QUANTIFICATION
OF IGNORANCE?

Hume has been answered based on a certain kind of model and a certain quantification of ignorance. A more radical skeptic might certainly question both assumptions. Bayes himself seems to have had qualms about his assumption of the uniform prior, buttressing it with an independent argument to the same conclusion. The assumption has been challenged in all sorts of ways. Why not a uniform prior on the square of the bias of the coin, and so forth? Can we do without a unique quantification of ignorance?

Ignorance is the opposite of knowledge. So, *an ignorance prior should be any prior that does not presume knowledge*. I might know the composition of the urn or the bias of the coin exactly. I might know less and still know something. I might know that the urn contains more black tickets than white or that the bias towards heads is greater than $\frac{1}{2}$. But suppose that I know no such thing.

Then my ignorance prior for a biased coin should put some positive probability on the true chance of being in any open interval between 0 and 1. Specification of an ignorance prior is not unique. There are lots of them. If you don't like calling these ignorance priors on the grounds that they may be sharply peaked, call them nondogmatic priors or *skeptical* priors, because these priors are quite in the spirit of ancient skepticism.[23]

What Laplace showed for the uniform prior holds for all skeptical priors for the biased coin. Given enough experience they lead a Bayesian to predict tomorrow using something close to the observed frequency. With *chance* 1, they lead the Bayesian to converge[24] to the true chances. The same is true for the biased die or for sampling from the "great urn of nature." *Skeptical priors defeat skepticism*.[25]

What about the dogmatist? Logic alone does not prevent one from being a dogmatist. Suppose, for example, that this person is convinced that the bias toward heads is greater than $\frac{1}{2}$, and has a uniform prior on it being between $\frac{1}{2}$ and 1. If the true bias is $\frac{1}{4}$, he will just never learn it. Nevertheless, he *believes* that he will learn the true chances, because he is *sure* that they are between $\frac{1}{2}$ and 1. You or I may not believe that he will learn the true chances, but *he* does. Third-person doubts are possible, but in a sense, first-person doubts are not. He believes what he believes. This point holds quite generally.[26] With *his degree of belief* 1, he will converge to the true chances.

The skeptical prior is not consistent with inductive skepticism at all. And even the dogmatic prior is inconsistent with first-person inductive skepticism. There is, however, *another big assumption* in the Bayes-Laplace model.

WHAT ABOUT THE EXISTENCE OF CHANCES?

The foregoing all takes place within a specific chance model. Perhaps, with Hume, we may believe that "there is no such thing as chance in the world."[27] We saw the answer to this enhanced skepticism in chapter 7. It was given by a great admirer of Hume, Bruno de Finetti.

Suppose there is a potentially infinite sequence of yes-no events. And suppose that you are a frequentist in the following weak sense:

For you, the only thing that matters for the probability of a finite outcome sequence of a given length is the relative frequency of successes in that sequence.

That is to say, for you, two sequences of the same length having the same relative frequencies have the same probability.

Then de Finetti *proves* that you *behave like Bayes*, with his chance model and some (not necessarily flat) prior. Furthermore, the prior is uniquely determined by your degrees of belief over the outcome sequences.[28]

De Finetti, like Hume, believes that there is no such thing as chance in the world and shows that we can have the virtues of Bayes' analysis without the baggage. If you are skeptical about the existence of chances, the chance model, and the prior over the chances, de Finetti shows how to get them all from your degrees of belief, provided that they satisfy the foregoing condition of exchangeability.

Furthermore, you must believe with probability 1 that a limiting relative frequency exists and that with repeated experience you will converge to it.[29] *If your degrees of belief are exchangeable, you cannot be an inductive skeptic.*

Your degrees of belief may not be exchangeable. There is no reason that they have to be. What then?

WHAT IF YOU DON'T HAVE EXCHANGEABLE DEGREES OF BELIEF?

Short of exchangeability there may be other symmetries in degree of belief, and such symmetries generally have inductive consequences. A slight weakening to allow for order effects gives Markov exchangeability. This can be extended to encompass different kinds of temporal and spatio-temporal order. More generally, symmetries in degrees of belief account for all kinds of analogical inference. De Finetti himself already initiated this line of thought in 1938,[30] and there are many subsequent developments.[31]

We saw in chapter 8 how ergodic theory, initially conceived of as a way to make sense of probabilities in statistical mechanics, can be viewed in an entirely different way as a far-reaching generalization of de Finetti's theorem.

Suppose that you have a space that encapsulates the problem that you are thinking about. You bring to this problem your degrees of belief: a probability measure on that space. Suppose that your degrees of belief have a *symmetry*—that they are invariant under a group of transformations of the space into itself. For example, the symmetry might be *exchangeability*—order doesn't matter. Then the transformations that swap sequences with the same frequencies but different orders in initial segments will leave your degrees of belief unchanged. Exchangeability is invariance of your probability under this group of transformations.

The transformations represent *your* conception of a repetition of an experiment.[32] Invariance means that the transformation (or group of transformations) leaves the probabilistic structure unchanged.

(We emphasize the subjective nature of the symmetry. You and I may have different conceptions of the repetition of an experiment, since we may have different symmetries in our degrees of belief.)

Starting at some point, x, in your probability space, you contemplate a series of repetitions of an experiment, x, Tx, TTx, . . . ,

$T^n x$. You keep track of the relative frequency of a property (of the points being in some measurable set, A).[33] Given the foregoing, you believe with probability 1 that the limiting relative frequency will exist![34, 35]

Notice that, *in this sense, you cannot be an inductive skeptic.* You cannot be a skeptic in the sense of Reichenbach. Such a skeptic doubts the existence of limiting relative frequencies. But you cannot doubt this, provided you are considering a sequence of repetitions of the *same experiment*, in *your* sense of same experiment. This is a consequence of the symmetry in your degrees of belief.

As emphasized at the onset, your probabilities and your conception of repetition of the same experiment are up to you. You and I may differ. *We may be skeptical about each other but not about ourselves.*

WHAT ABOUT THE PREDICATES USED TO DESCRIBE THE WORLD?

In *Fact, Fiction and Forecast*[36] (1955), the philosopher Nelson Goodman concocts the "grue" hypothesis to show the hopelessness of purely syntactical theories of confirmation. The predicate grue applies to all things examined before t just in case they are green and to all other things just in case they are blue. At time t our evidence is that all previously examined emeralds have been green. Then, providing we know what time it is, our evidence is also to the effect that all previously examined emeralds have been grue. The form of the syntax is the same: that is the hypotheses are of the form "all Xs are Ys" and the evidence for them is of the form "all observed Xs have been Ys." But we do not take the two hypotheses "all emeralds are green" and "all emeralds are grue" to be equally well confirmed on the evidence.

Goodman concludes that regularities in terms of some kinds of predicates are projectible into the future, but regularities in terms of others are not. He sees the problem of characterizing the class of projectible predicates as a *new* riddle of induction, and it received a lot of discussion in the second half of the twentieth century.

Let us ask how Goodman's example looks from the viewpoint of de Finetti. De Finetti is at least as radical as Goodman in questioning the basis of the received categories (Bruno de Finetti, "Sur la condition d'équivalence partielle," *Actualités scientifiques et Industrielles* 739 (1938), translated as "On the Condition of Partial Exchangeability" by P. Benacerraf and R. Jeffrey in *Studies in Inductive Logic and Probability II*, ed. R. Jeffrey (Berkeley: University of California Press, 1980): 194):

> What are sometimes called repetitions or trials of the same event are for us so many distinct events. In general they will have common or symmetric characteristics which make it natural to attribute equal probabilities to them, but there is no a priori reason which prevents us in principle from attributing to the events $E_1 \ldots E_n$ distinct and absolutely arbitrary probabilities $p_1 \ldots p_n$. In principle there is no difference for us between this case and that of n events which exhibit no similarities; the similarity which suggests the name "trials of the same event" (we would say "of the same phenomenon") is not intrinsic: at the very most, its importance lies in the influence it may have on our psychological judgment . . . .

Judgments of similarity are, for de Finetti, subjective, and they are embodied in one's subjective probabilities. Suppose a coin of unknown bias is about to be flipped and you are to be presented with the results in an unusual way. For the first 100 trials there is a random variable that takes the value 1 if heads, 0 otherwise. For the subsequent trials there are random variables that take the value 0 if heads, 1 otherwise.

If you have ordinary beliefs about coin flipping, this "Goodmanized" sequence of random variables will just not be exchangeable for you. But a sequence of random variables that uniformly gives 0 for heads and 1 for tails will be exchangeable for you. Logic, however, in no way requires you to have ordinary beliefs. Someone could consistently have degrees of belief according to which the Goodmanized sequence was exchangeable. For that person, the conventional sequence would not be exchangeable. Both you and he would believe that the future was like the past, but in different ways.

Goodman's concerns have already been addressed by de Finetti.[37] Projectibility is captured in a subjective Bayesian setting by *exchangeability* or generalizations thereof.*

WHAT ABOUT UNCERTAIN EVIDENCE?

So far, the envisioned learning experiences have been modeled as conditioning on the evidence, which comes nicely packaged as a proposition.[38] A more radical skeptic may well call even this into question. This is the stance taken in Richard Jeffrey's "Radical Probabilism."[39] Must a radical probabilist perforce be a radical inductive skeptic?

Richard Jeffrey

Suppose that one learns through some sort of black-box interaction that updates one's probabilities (figure 10.1). Then we need some way of distinguishing interactions that are viewed as learning experiences and those that are viewed as mindworms, brainwashing,

---

* We include generalizations because Goodman talks about projecting patterns in past data into the future. By patterns, he might just mean regularities. But if we take patterns seriously, we will move to Markov exchangeability and other forms of partial exchangeability. What Goodman called projectibility takes various forms, and these different forms of projectibility amount to different symmetries in one's degrees of belief.

Figure 10.1. A cat goes through a black-box learning experience.

drug-induced hallucinations, the Sirens singing to Ulysses, and so on. A plausible candidate is diachronic coherence.[40, 41]

If one contemplates a sequence of such experiences, stretching off into the future (as in figure 10.2), and regards them as learning experiences, coherence requires that they form a martingale in your degrees of belief.[42] The sequence of revised probabilities forms a martingale. That means that the martingale convergence theorem comes into play. You must believe that your degrees of belief will converge.

As an added twist, suppose one is even skeptical of the mathematical idealizations of standard probability theory. The standard martingale convergence theorem uses the framework of Kolmogorov, which we discussed in chapter 5, and that framework uses countably infinite additivity. We have seen that de Finetti himself was skeptical about this idealization. But one who is skeptical of countable additivity, or

Figure 10.2. A cat contemplates a sequence of black-box learning experiences.

continuity, used by Kolmogorov need not worry. This can all be done with finitely additive martingales.[43]

By this point, the skeptic has called *almost everything* into question. Nevertheless, although we cannot say anything about what we believe our degrees of belief will converge to, we must believe that they will converge. Even in this austere setting, one cannot be a complete inductive skeptic.

SUMMING UP

Hume remarks that it is *psychologically* impossible to be a consistent skeptic:

> since reason is incapable of dispelling these clouds, nature herself suffices to that purpose . . . .

This is, in broad outline, psychologically correct, even though—as we saw in chapter 2—human psychology may exhibit systematic quirks.

But, as Hume maintained and Russell emphasized, it is nevertheless *logically* possible to be a consistent skeptic. Even when we have
tried to assume as little as possible, we have still assumed something.
One is not logically compelled to believe that one will face a sequence*
of learning experiments. One may not be coherent or believe that one
will remain coherent in the future. One need not believe that there
will be a future. Absolute skepticism is unanswerable.

But short of absolute skepticism, there are various grades of inductive skepticism, differing in what the skeptic brings to the table and
what he calls into doubt. Some kinds of skeptics may call into question things to which they are implicitly committed. In such a case,
reason *is* capable of dispelling doubts.

In one setting, Bayes, Laplace, and their followers solved the problem of induction. With fewer assumptions, in a setting more congenial to Hume, de Finetti and his heirs solved Hume's problem. It
is remarkable the extent to which the logic of coherent belief itself
constrains inductive skepticism.

---

*An infinite sequence is, of course, fantasy. But remember the finite forms of de Finetti.

APPENDIX: PROBABILITY TUTORIAL

We are assuming that our reader has taken a first undergraduate course in probability or statistics. The marvelous books by Freedman, Pisani, and Purves* or Feller[†] have more than we need. In case this was years ago, we include a brief tutorial covering notation of the basic model, sample space, and summation notation; an example of a nontransitivity paradox; basic facts such as sum rule, independence and product rule, conditional probability, Bayes' theorem, law of total probability; a discussion of random variables and expectation; and, finally, an introduction to conditional expectation and martingales.

NOTATION: WRITING THINGS DOWN

Writing out probability results uses notation: sets and summation. This section contains a brief tutorial. A first building block is the notion of a finite set; for example,

$$\mathcal{X} = \{\text{Persi, Brian, Bill}\}$$

is a set of three names.

$$\mathcal{X} = \{2, 3, 5, 7, 11, 13\}$$

---

*D. Freedman, R. Pisani, and R. Purves, *Statistics*, 4th ed. (New York: Norton, 2007).

[†]W. Feller, *An Introduction to Probability Theory and its Applications*, 3d ed. (New York: Wiley, 2008).

is the set of the first 6 prime numbers. Often we don't need to specify what's in the set, but it's still useful to have placeholders. We write things like let $\mathcal{X} = \{x_1, x_2, \ldots, x_N\}$ be a set with N elements. In the preceding examples, $N = 3$ and $x_1 = $ Persi, $x_2 = $ Brian, $x_3 = $ Bill or $N = 6$, $x_1 = 2$, and so on.

The second building block is the notion of probability measure or distribution. A *probability* on a finite set is a collection of numbers $P(x_1), P(x_2), \ldots, P(x_N)$ with $P(x)$ positive—we write $P(x) \geq 0$ and in fact $P(x) = 0$ is allowed—and with total sum 1, that is, $P(x_1) + P(x_2) + \cdots + P(x_N) = 1$. For example, $P(\text{Persi}) = P(\text{Brian}) = P(\text{Bill}) = \frac{1}{3}$ or $P(\text{Persi}) = \frac{1}{10}$, $P(\text{Brian}) = \frac{1}{5}$, $P(\text{Bill}) = \frac{7}{10}$.

From these two pieces, the basic mathematical question of probability is easy to state.

BASIC QUESTION OF PROBABILITY

Given $\mathcal{X} = \{x_1, x_2, \ldots, x_N\}$, a set with N elements; $P(x_1), P(x_2), \ldots, P(x_N)$, a probability on $\mathcal{X}$; and a subset A of $\mathcal{X}$, calculate or approximate $P(A)$, the sum of $P(x)$ for x in A.

Example: If $\mathcal{X} = \{$Persi, Brian, Bill$\}$ and $P(\text{Persi}) = P(\text{Brian}) = P(\text{Bill}) = \frac{1}{3}$ and $A = \{$Brian, Bill$\}$, $P(A) = P(\text{Brian}) + P(\text{Bill}) = \frac{2}{3}$.

Example: If $\mathcal{X} = \{1, 2, 3, 4, 5, 6, 7, 8, 9, 10\}$, $P(x) = \frac{1}{10}$ for all x, and $A = \{2, 3, 5, 7\}$ is the set of primes in $\mathcal{X}$, $P(A) = \frac{4}{10}$.

In this *mathematical* version of probability, someone tells us $\mathcal{X}$, $P(x)$, and A and our job is to calculate $P(A)$. This can be difficult if N is large or $P(x)$ is given in an indirect way, or A is a complicated set; see the birthday problem example in chapter 1.

There is some notation that is very useful but off-putting if you haven't seen it before: this is summation notation. If $P(x)$ is a probability on $\mathcal{X}$ and A is a subset, we write

$$P(A) = \sum_{x \in A} P(x).$$

The right-hand side is read, the sum of $P(x)$ for x in A. Thus, for the first example, with A the names in $\mathcal{X}$ that start with B,

$$\sum_{x \in A} P(x) = P(\text{Brian}) + P(\text{Bill}) = \frac{2}{3}.$$

We call the specification of the sample space $\mathcal{X}$ and the probability $P(x)$ *the standard model*.

AN EXAMPLE: A NONTRANSITIVITY PARADOX

Here is an example of the basic setup with a surprising conclusion. We will build an example where A beats B, B beats C, but C beats A. Of course, relations such as "beats" are sometimes nontransitive: if A loves B and B loves C, often A does not love C. The classic rock-paper-scissors game is another well-known example. Still, we bet you'll find the following example surprising!

Begin with a classical magic square:

4	3	8
9	5	1
2	7	6

All the rows and all the columns and both diagonals sum to 15. Make a game from this by considering the columns as three piles of cards:

Pile I	Pile II	Pile III
$\{4, 9, 2\}$	$\{3, 5, 7\}$	$\{8, 1, 6\}$

If it helps, remove the $1, 2, \ldots, 9$ of hearts from a deck of cards, counting the ace as 1, and form the three piles shown. The game is as follows: You pick one of the three piles and we pick one of the remaining piles; each of us shuffles our piles (no cheating), and turns up the top card; high card wins. We claim that whatever pile you choose, we can pick a pile that beats you! Suppose you pick pile I $\leftrightarrow \{4, 9, 2\}$. We will pick pile II $\leftrightarrow \{3, 5, 7\}$ and claim that we win with probability

$\frac{5}{9}$. Remember, this is an exercise to teach simple probability calculations: What is $\mathcal{X}$, the sample space? What is $P(x)$, the underlying probability? What is A, the set of interest? And finally, what is $P(A)$?

At the moment, the game involves piles I and II. When these are mixed and the two top cards are turned up, the possible outcomes are

$$\mathcal{X} = \{(4,3), (4,5), (4,7), (9,3), (9,5), (9,7), (2,3), (2,5), (2,7)\}.$$

Here, for example, $(4,3)$ means the 4 turns up on top of pile I and the 3 turns up on top of pile II. Thus $\mathcal{X}$ has 9 elements. Our fair-mixing assumption shows that all 9 outcomes have an equal chance. Thus $P(x) = \frac{1}{9}$. We want to calculate the chance, say, that player II (that's us) wins. This corresponds to

$$A = \{(4,5), (4,7), (2,3), (2,5), (2,7)\}.$$

From this,

$$P(\text{pile II beats pile I}) = P(A) = \frac{5}{9}.$$

Please verify

$$P(\text{pile III beats pile II}) = \frac{5}{9}, \quad P(\text{pile I beats pile III}) = \frac{5}{9}.$$

You can also check that the piles formed from the three rows of the magic square give nontransitivity.

There are many variations: when we were kids, we observed that the standard 4×4 magic square doesn't give nontransitive piles, but the standard construction of odd-order squares does result in nontransitive piles. See the Wikipedia entry on nontransitive dice, https://en.wikipedia.org/wiki/Nontransitive_dice, for more.

BASIC FACTS: THE RULES OF THE GAME

Going back to the general case, given any $\mathcal{X}, P(x)$, there are a handful of simple consequences of the basic model that simplify calculations:

- the sum rule for disjoint events

- the product rule for independent events

- conditional probability and Bayes' theorem

- the law of total probability

Each of these is a simple little theorem starting from the standard model. Of course, learning to use basic tools fluidly in new situations takes practice. Throughout, $\mathcal{X}$ is a finite set and $P(x)$ is a probability on $\mathcal{X}$.

SUM RULE

If A and B are subsets of $\mathcal{X}$ with no element in common, then, with $A \cup B = \{x \text{ in } A \text{ or } B\}$,

$$P(A \cup B) = P(A) + P(B).$$

Example: Let $\mathcal{X}$ correspond to the experiment of shuffling a normal deck of 52 cards and turning up the top card. Thus

$$\mathcal{X} = \{AC, 2C, \ldots, KC, AH, \ldots, KD\}$$

and $P(x) = \frac{1}{52}$ for all x, where AC stands for the ace of clubs for example and KD stands for the king of diamonds. Let A correspond to the event that an ace turns up. Thus $A = \{AC, AH, AS, AD\}$. Clearly

$$P(A) = \frac{1}{13}.$$

Let B correspond to the event that a 2 turns up; so $B = \{2C, 2H, 2S, 2D\}$ and $P(B) = \frac{1}{13}$ again. Then $A \cup B$ corresponds to the event that an ace or a two turns up. Since clearly A and B have no element in common,

$$P(A \cup B) = P(A) + P(B) = \frac{2}{13}.$$

On the other hand, if B had corresponded to the event that a club turns up,

$$B = \{AC, 2C, 3C, 4C, 5C, 6C, 7C, 8C, 9C, 10C, JC, QC, KC\}$$

with $P(B) = \frac{1}{4}$, then $P(A \cup B) \neq P(A) + P(B)$.

Exercise: For any subsets A and B in any probability space, verify the rule

$$P(A \cup B) = P(A) + P(B) - P(A \cap B),$$

where $A \cap B$, the intersection of A and B, is the set of common elements. Use this formula to show that

$$P(\text{an ace or a club}) = \frac{1}{13} + \frac{1}{4} - \frac{1}{52} = \frac{16}{52}.$$

Exercise: With a well-shuffled deck of 52 cards, let A be the event that an ace turns up on top. Let B be the event that a club turns up as the second card. What is $P(A \cap B)$? Can you explain? Can you generalize?

INDEPENDENCE AND THE PRODUCT RULE

A and B are independent if

$$P(A \cap B) = P(A)P(B).$$

Note that independence depends both on A, B, and the probability P.

Example: In the turning-up-the-top-card example, if A corresponds to the event where an ace turns up and B corresponds to the event where a club turns up, then $A \cap B = \{\text{AC}\}$ and

$$P(A \cap B) = \frac{1}{52} = \frac{1}{11} \cdot \frac{1}{13} = P(A)P(B).$$

CONDITIONAL PROBABILITY

We define conditional probability for subsets A and B (with $P(B) > 0$) as

$$P(A \mid B) = \frac{P(A \cap B)}{P(B)}.$$

The left-hand side is read as the probability of A given that B has occurred. It is defined by taking the underlying probability P, restricting it to B—that's $P(A \cap B)$—and then renormalizing so that the total sum is 1.

Example: In our turning-up-the-top-card example, if B corresponds to the event where a club turns up and A corresponds to the event where a card higher than 7 turns up, then

$$P(A \mid B) = \frac{7}{13}.$$

Observe that if A and B are independent $P(A \mid B) = P(A)$. Can you construct an example of $\mathcal{X}, P(x), A, B$ where $P(A \mid B) = P(A)$ but A and B are *not* independent?

BAYES' THEOREM

If A and B are any subsets, both with positive probability, then

$$P(A \mid B) = \frac{P(B \mid A)P(A)}{P(B)}.$$

This simple formula follows at once from the definitions. Replacing $P(A \mid B)$ by $P(A \cap B)/P(B)$ and $P(B \mid A)$ by $P(B \cap A)/P(A)$ gives

$$\frac{P(A \cap B)}{P(B)} \overset{?}{=} \frac{P(B \cap A)}{P(A)} \frac{P(A)}{P(B)}.$$

Canceling equal terms on both sides, the equality follows from $P(A \cap B) = P(B \cap A)$.

Example: In the turning-up-the-top-card example, with A corresponding to the event of a "high card" (7 or above) and B corresponding to the event of a club card, $P(A \mid B) = \frac{7}{13}$, $P(B \mid A) = \frac{1}{4}$, $P(A) = \frac{7}{13}$, $P(B) = \frac{1}{4}$, and the result checks.

Exercise: There are 3 urns, each containing 2 balls. Urn I contains 2 white balls, urn II contains 1 red and 1 white ball, and urn III contains 2 red balls. An urn is chosen at random, with probability $\frac{1}{3}$, and a ball is chosen at random from that urn, with probability $\frac{1}{2}$. You see the color of the chosen ball is red. What is the chance that the second ball in the urn is red?

THE LAW OF TOTAL PROBABILITY

This is a simple, useful consequence of the preceding definitions. Let $B_1, B_2, \ldots, B_k$ be a decomposition of $\mathcal{X}$ into disjoint subsets with $P(B_i) > 0$ for all i. Then, for any set A,

$$P(A) = \sum_{i=1}^{k} P(A \mid B_i)P(B_i).$$

Example: Return to the exercise with the three urns. Let $A = $ {a red ball is chosen}. Let $B_i = $ {urn i is chosen} for $i = $ I, II, III. Then $P(B_i) = \frac{1}{3}$, $P(A \mid B_1) = 0$, $P(A \mid B_2) = \frac{1}{2}$, and $P(A \mid B_3) = 1$, so

$$P(A) = 0 \cdot \frac{1}{3} + \frac{1}{2} \cdot \frac{1}{3} + 1 \cdot \frac{1}{3} = \frac{1}{2}.$$

In retrospect, $\frac{1}{2}$ seems obvious "by symmetry." Many things seem obvious in hindsight.

RANDOM VARIABLES AND EXPECTATION

There is a simple extension of the basic model that is extremely useful. Let $\mathcal{X}$ be a finite set and $P(x)$ a probability on $\mathcal{X}$. A *random variable* is a function $X(x)$ assigning numbers to points of $\mathcal{X}$. Its *expectation* is the average with points weighted by $P(x)$,

$$E(X) = \sum_{x} X(x)P(x).$$

Example: In the previous turn-up-the-top-card example, $\mathcal{X} = $ {AC, 2C, \ldots, KD} and $P(x) = \frac{1}{52}$. Let $X(x)$ be the value of card x, so $X(7C) = 7$, $X(AC) = 1$, $X(KD) = 13$, and so on. Then

$$E(X) = \frac{1}{52}(1 + 2 + \cdots + 13 + 1 + 2 + \cdots + 13 + \cdots$$
$$+ 1 + 2 + \cdots + 13)$$
$$= \frac{1}{13}(1 + 2 + \cdots + 13) = 7.$$

A simple consequence of the preceding definitions is the most useful *linearity* property. If X and Y are random variables, then

$$E(X+Y)=E(X)+E(Y).$$

This is reminiscent of the sum rule, but linearity of expectation holds for *any* random variables: no disjointness or independence is needed.

Example: Consider a card-guessing experiment with an ordinary deck of 52 cards. After shuffling, a "guesser" tries to guess the current top card of the deck. After each guess, the current top card is shown and discarded. How many correct guesses are expected, going through the deck?

Answer: The answer depends on how the guesser uses the information as the experiment proceeds. Consider four scenarios:

1. The guesser pays no attention to the information and always guesses AC.

2. The guesser pays no attention to the information and guesses a randomly chosen card each time.

3. The guesser pays attention to the information and always guesses a card known to still be in the deck.

4. The guesser chooses AC first and afterward always guesses the first card that was shown.

These four strategies could be called Idiotic, Random, Greedy, and Worst Case. With all this specified, what are the four expected values? For each, let X_i be the random variable which is 1 if guess i is correct and 0 otherwise. The total number of correct guesses is

$$S=X_1+X_2+\cdots+X_{52}$$

and we are asked to determine $E(S)$.

Case 1. Idiotic: By linearity, $E(S)=E(X_1)+E(X_2)+\cdots+E(X_{52})$. Clearly, $E(X_1)=\frac{1}{52}$, and similarly $E(X_2)=E(X_3)=\cdots=E(X_{52})=\frac{1}{52}$. Thus

$$E(S)=\frac{1}{52}+\cdots+\frac{1}{52}=1.$$

Case 2. Random: By the same reasoning,

$$E(S) = 1.$$

For those who know what it means, the *variance* in case 1 is 0, while the variance in case 2 is $1 - \frac{1}{52}$.

Case 3. Greedy: Now, $E(X_1) = \frac{1}{52}, E(X_2) = \frac{1}{51}, E(X_3) = \frac{1}{50}, \ldots, E(X_{52})$ $= 1$ so

$$E(S) = 1 + \frac{1}{2} + \cdots + \frac{1}{52} \doteq 4.5.$$

Case 4. Worst Case: Here, $E(X_1) = \frac{1}{52}$ but $E(X_i) = 0$ for all i greater than 1, so

$$E(S) = \frac{1}{52}.$$

In these, and many other calculations, linearity makes hard calculations easy.

Exercise: Let $X_i = i$ if guess i is correct, zero if not. What is $E(S)$ for strategies 1, 2, 3, and 4?

CONDITIONAL EXPECTATION AND MARTINGALES

Conditional probability, random variables, and expectation can be combined. If X and Y are random variables, define the *conditional expectation* of Y given $X = x$ as

$$E(Y \mid X = x) = \sum_z Y(z) P(z \mid X = x).$$

On the right, $P(z \mid X = x) = P(z)/P(B)$ if z is in B and zero if z is not in B, where $B = \{y : X(y) = x\}$.

Example: In the turn-up-the-top-card example, if Y is the value of the top card and X is 1 or 0 as the top card is 7 or above (or not), $P(z \mid X = 1) = \frac{1}{7}$ if z is a card having value 7 or above and 0 otherwise. Then

$$E(Y \mid X = 1) = \frac{1}{7}(7 + 8 + 9 + 10 + 11 + 12) = \frac{57}{7} \doteq 8.14.$$

Similarly

$$E(Y \mid X = 0) = \frac{1}{6}(1 + 2 + 3 + 4 + 5 + 6) = \frac{7}{2} \doteq 3.5.$$

Conditional expectation is still linear, as a function of Y. If X and Y are independent,

$$E(Y \mid X = x) = E(Y).$$

With these definitions, we can take a step from elementary to modern probability. Let $X_1, X_2, \ldots, X_n$ be random variables. They form a *martingale* if, for each i,

$$E(X_i \mid X_1 = x_1 \cdots X_{i-1} = x_{i-1}) = x_{i-1}.$$

Thus, the future expectation depends on the past only through the present. It is outside our scope to develop this theory but it *is* achievable within the confines of elementary probability; see Grinstead and Snell* or Lange†.

AN EXAMPLE: PÓLYA'S URN

The next example combines most all of the preceding ingredients, from basic probability through martingales. We think anyone can follow along. It illustrates how people who work with probabilities talk and think.

The start is simple enough: picture an urn containing two balls, one labeled 0 and the other labeled 1. The rules are simple, too. Each time a ball is chosen at random (uniformly) from the urn, it is then replaced along with another ball bearing the same label. Thus, after two rounds, here are the possible evolutions:

---

*C. Grinstead and J. L. Snell (2012), *Introduction to Probability*, 2d rev. ed. (Providence, RI: American Mathematical Society, 2012).

†K. Lange, *Applied Probability* (Berlin: Springer, 2010).

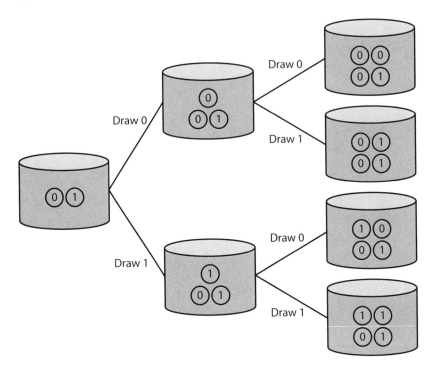

Pólya and Eggenberger* introduced this with a health-related story involving contagion: a population evolves and the outcomes effect the chances of future outcomes. Notice that it would be a chore to write down the sample space and underlying probability model for Pólya's urn. We *could* do it: after two draws, there are four outcomes

$$\mathcal{X} = \{00, 01, 10, 11\} \text{ with } P(00) = \frac{1}{3}, \quad P(01) = P(10) = \frac{1}{6}, \quad P(11) = \frac{1}{3}.$$

After three draws there are eight outcomes, with e.g., $P(000) = \frac{1}{4}$. After n draws there are 2^n possible outcomes, with, for example, $P(0\ldots0) = \frac{1}{(n+1)}$. The actual probabilities can be written out, but we don't have to do it to go forward.

*F. Eggenberger and G. Pólya, "Über die Statistik verketteter Vorgänge," *Journal of Applied Mathematics and Mechanics*, 3 no. 4 (1923): 279–89.

Example: Let X_n be the proportion of balls labeled 1 in the urn at time n. Thus $X_0 = \frac{1}{2}$, and if the sequence of draws starts with 010,

$$X_0 = \frac{1}{2}, \quad X_1 = \frac{1}{3}, \quad X_2 = \frac{1}{2}, \quad X_3 = \frac{2}{5}, \ldots.$$

We claim the sequence of random variables $X_0, X_1, X_2, \ldots$ forms a martingale. This is easy to see: after n draws, the next proportion only depends on the last; the number of balls in the urn at time n is $n + 2$, and

$$E(X_{n+1} \mid X_1, X_2, \ldots, X_n) = E(X_{n+1} \mid X_n) = X_n.$$

Exercise: Check this last computation in your own way when $n = 3$. Show that, if there are initially a 0s and b 1s in the urn at time 0, the X_n sequence forms a martingale.

You may well ask "so what?" Well, martingales have a host of properties. To begin with, X_n converges to a limit as n gets large. (This is not so easy to see: why doesn't it oscillate?) There is the optional sampling property—stopping at a random time T, $E(X_T) = E(X_0) = \frac{1}{2}$—central limit theorems, easy bounds for fluctuations such as the Azuma–Hoeffding inequality, ... These topics are the warp and woof of a first graduate course in probability; they have also made a mark in computer science. Indeed, our colleague Don Knuth's magisterial *The Art of Computer Programming Volume 4B* (Boston: Addison-Wesley Professional, 2015) contains some novel properties of both martingales and Pólya urns.

FROM DISCRETE TO CONTINUOUS AND BEYOND

So far, we have developed probability with finite spaces. It is straightforward to generalize to continuous spaces and much beyond. If $f(x)$ is a function of a real variable x with $f(x) \geq 0$, $\int_{-\infty}^{\infty} f(x)dx = 1$, and A is a subset, define

$$P(A) = \int_A f(x)\, dx.$$

Of course, sometimes the integral on the right side has to be carefully defined, and this can be the stuff of a graduate course in real analysis. The big picture hardly changes; see chapter 5 for Kolmogorov's axiomatic version. All the preceding properties hold for this version of P.

Anyway, anything that anyone can observe in the real world is measurable only to finitely many decimals. If you master discrete probability, you are in very good shape.

THE COMPUTER IS HERE!

There has been a sea change for the mathematics of probability: computer simulation is replacing theorems. Almost any of the problems discussed here can be "solved" by simply using the output of computer-generated random numbers and repeating the experiment a few thousand (or a few million) times. This takes seconds to do, depending on programming time, and gives results for complex problems that math can't touch. Take our first example, the nontransitive three-piles problem. It is a simple matter to take pile I, with $\{4, 9, 2\}$, and pile II, with $\{3, 5, 7\}$, and ask the computer to repeatedly choose randomly from each and keep track of who wins.

There are still things to think about. How do we know that the random number generator is really random? (And what does that mean anyway? See chapter 8.) For more complicated problems, how many repetions suffice? It can also evolve into a research problem to generate random instances: how do you tell the computer to shuffle cards? All of this is for the good, opening up new research areas. Our two favorite sources for this Monte Carlo approach to probability are the classic Hammersley and Handscomb[*] and more recent Liu.[†]

It is wise to remember that the real world is different than a simulated world. There may be key features to observe that explain what

[*]J. M. Hammersley and D. C. Handscomb, *Monte Carlo Methods* (Berlin: Springer, 1964).

[†]J. S. Liu, *Monte Carlo Strategies in Scientific Computing* (Berlin: Springer, 2001).

you are after, but these are absent from your simulations. Finally, there *are* problems—lots of them—where we don't know how to use simulations. How many times should a deck of cards be shuffled? We can simulate shuffling, but with 10^{68} or so possible outcomes we will never get a good picture or useful answer.

NOTES

CHAPTER 1

1. Echoed by Cicero in *De Natura*.

2. A superb history of early probability is in James Franklin's *The Science of Conjecture: Evidence and Probability before Pascal* (Baltimore: Johns Hopkins University Press, 2002). Franklin examines every scrap of evidence we have, from the Talmud, early Roman law, and insurance over many ethnicities. He makes it clear that people had all sorts of thoughts about chance, but not a single quantitative aspect surfaces.

3. The same issues come up in measuring any basic quantity, for example, the weight of the standard kilogram or the frequency of light. Careful discussion is the domain of measurement theory. For an extensive discussion, see D. H. Krantz, R. D. Luce, P. Suppes, and A. Tversky, *Foundations of Measurement, Vol. I* (1971), *Vol II* (1989), *Vol III* (1990), (New York: Academic Press). For an illuminating discussion of how the Bureau of Standards actually measures the standard kilogram, see D. Freedman, R. Pisani, and R. Purves, *Statistics*, 4th ed. (New York: W. W. Norton, 2007).

4. For discussion, see S. Stigler, *Seven Pillars of Statistical Wisdom* (Cambridge, MA: Harvard University Press, 2016).

5. This kind of virtuous circularity appears throughout science. For an illustration in a much richer setting, see George E. Smith, "Closing the Loop: Testing Newtonian Gravity, Then and Now," in *Newton and Empiricism*, ed. Zvi Beiner and Eric Schliesser (Oxford: Oxford University Press, 2014): 262–351.

6. Written about 1564 but published only posthumously. See O. Ore, *Cardano, the Gambling Scholar* (Princeton: Princeton University Press, 1953), for translation and commentary.

7. Our chapter 7. There are a lot of connections between the early gambling literature and the foundations of probability. See D. Bellhouse, "The Role of Roguery in the History of Probability," *Statistical Science* 8 (1993): 410–20.

8. See appendix 1.

9. Fermat sees this clearly, but Pascal seems to have either made a mistake or misidentified the problem. See appendix 1.

10. Known in India since the second century BCE. For more of the history, see A. Edwards, *Pascal's Arithmetical Triangle: The Story of a Mathematical Idea* (Baltimore: Johns Hopkins University Press, 2002).

11. For more on Huygens, see S. Stigler, "Chance Is 350 Years Old," *Chance* 20 (2007): 33–36.

12. Isaac Newton had a copy and made notes, which can be found in D. T. Whiteside, ed., *The Mathematical Papers of Isaac Newton, Volume 1* (Cambridge: Cambridge University Press, 1967). Thanks to Stephen Stigler for the reference.

13. Incidentally, Newton, one of the great mathematicians of all time, had poor probabilistic skills. See S. Stigler, "Issac Newton as a Probabilist," *Statistical Science* 21 (2004): 400–403.

14. Translated with extensive commentary by Edith Dudley Sylla as *The Art of Conjecturing, Together with a Letter to a Friend on Sets in Court Tennis* (Baltimore: Johns Hopkins University Press, 2006).

15. T. Gilovitch, R. Vallone, and A. Tversky, "The Hot Hand in Basketball: On the Misperception of Random," *Cognitive Psychology* 17 (1985): 295–314.

16. Joseph Keller, "The Probability of Heads," *American Mathematical Monthly* 93 (1986): 191–97.

17. P. Diaconis, S. Holmes, and R. Montgomery, "Dynamical Bias in the Coin Toss," *Siam Review* 49 (2007): 211–35.

18. D. Bayer and P. Diaconis, "Tracking the Dovetail Shuffle to Its Lair," *Annals of Applied Probability* 2 (1992): 294–313.

19. Quoted from the translation of Sylla, 327.

20. P. Diaconis and F. Mosteller, "Methods for Studying Coincidences," *Journal of the American Statistical Association* 84 (1989): 853–61.

21. P. Diaconis and S. Holmes, "A Bayesian Peek into Feller, Volume I," *Sankhya* 64 (2002): 820–41.

CHAPTER 2

1. Leibniz thought that, given the evidence, these probabilities would be determined by logic. This was the position taken by J. M. Keynes in the early twentieth century in his *Treatise on Probability* (London: Macmillan, 1921). It was effectively criticized by Frank Ramsey, who put forward a judgmental (personalist, subjective) interpretation in 1926, "Truth and Probability," in Frank Ramsey, *The Foundations of Mathematics and Other Logical Essays*, Ch. VII, ed. by R. B. Braithwaite (London: Kegan, Paul, Trench, Trubner & Co., New York: Harcourt, Brace and Company, 1931): 156–98.

2. Paul W. Rhode and Koleman Strumpf, "The Long History of Political Betting Markets: An International Perspective," *Oxford Handbooks Online* (Oxford: Oxford University Press, 2013).

3. There are many further instances of probability in the real world that can serve as substitutes for betting. David Aldous' marvelous online book *On Chance and Uncertainty* is our top recommendation. He has many further pointers.

4. L. J. Savage, *The Foundations of Statistics* (New York: Wiley, 1954).

5. For those who are concerned about infinite spaces, there is a completely analogous argument for countable additivity using a countably infinite number of bets. See E. Adams, "On Rational Betting Systems," *Archiv für mathematische Logik und Grundlagenforschung* 6 (1962): 7–29. Is it legitimate to use a countable number of bets? This is controversial. De Finetti thought not. For further discussion, see B. Skyrms, "Strict Coherence, Sigma Coherence and the Metaphysics of Quantity," *Philosophical Studies* 77 (1995): 39–55. Reprinted in Skyrms, *From Zeno to Arbitrage* (Oxford: Oxford University Press, 2012).

6. Conditional probabilities explain much of the pragmatics of indicative conditional constructions—If Mary is lying, then Samantha is guilty—in natural language. See E. Adams, *The Logic of Conditionals: An Application of Probability to Deductive Logic*, Synthese Library 86

(Dordrecht, Holland: Reidel, 1975). You may consider whether the ensuing discussion provides a rationale.

7. Communicated by Paul Teller, "Conditionalization and Observation," *Synthese* 26 (1973): 218–58.

8. The case where $P(A|e) < P_e(A)$ is similar.

9. *Let's work it out:* Suppose *e* doesn't happen. Then only the bets from the first day count. The bookie paid out fair prices for the 3 bets, respectively:

1. P (A and e),
2. $(1 - P(e))$ $P(A|e) = P(A|e) - P(A$ and $e)$,
3. $\delta P(e)$.

He loses bets 1 and 3 but wins bet 2 and collects $P(A|e)$.

Adding it up, the bookie loses $\delta P(e)$.

Suppose *e* does happen. Then the second transaction is made. We buy back the bet on *A* from the bookie. He gets

$$P(A|e) - \delta.$$

It now makes no difference whether *A* happens or not. But the bookie has won bet 3 and so gets

$$\delta.$$

Adding it up, the bookie loses $\delta P(e)$, as before.

10. David Freedman and Roger Purves, *Annals of Mathematical Statistics* 40 (1969): 1177–86.

11. Richard Jeffrey (1965), *The Logic of Decision*, 3rd rev. ed. (New York: McGraw-Hill; Chicago: University of Chicago Press, 1983).

12. See P. Diaconis and S. Zabell, "Updating Subjective Probability," *Journal of the American Statistical Association* 77 (1982): 822–30.

13. B. Skyrms, "Dynamic Coherence and Probability Kinematics," *Philosophy of Science* 54 (1987): 1–20. Reprinted in B. Skyrms, *From Zeno to Arbitrage* (New York and London: Oxford University Press, 2012).

14. See B. Skyrms, *The Dynamics of Rational Deliberation* (Cambridge, MA: Harvard University Press, 1990) and "Diachronic Coherence and Radical Probabilism," *Philosophy of Science* 73 (2006): 959–68. Reprinted in B. Skyrms. *From Zeno to Arbitrage* (New York and London: Oxford University Press, 2012).

15. B. de Finetti (1974), *Theory of Probability*, tr. A. Machi and A. Smith (New York: Wiley, 1999).

16. There is another motivation as well, based on incentives to reveal one's true probabilities.

17. Also known as mixtures, or convex combinations.

18. See A. H. Murphy and R. M. Winkler, "Probability Forecasting in Meteorology," *Journal of the American Statistical Association* (1984): 489–500, and L. J. Savage, "Elicitation of Personal Probabilities and Expectations," *Journal of the American Statistical Association* 66 (1971): 783–801, respectively.

19. This is called convexity.

20. "Correspondence of Nicholas Bernoulli concerning the St. Petersburg Game," tr. from Richard J. Pullskamp, *Die Werke von Jacob Bernoulli Band 3 K9*, Xavier University (2013). Posted at http://cerebro.xu.edu/math/Sources/NBernoulli/correspondence_petersburg_game.pdf.

21. On either of these suggestions, one can recreate the problem, but postulating prizes that grow fast enough as pointed out by Karl Menger in 1934: K. Menger, "The Role of Uncertainty in Economics," in Martin Shubik, ed., *Essays in Mathematical Economics in Honor of Oskar Morgenstern* (Princeton: Princeton University Press, 1934; 1966).

22. D. Bernoulli (1783), tr. L. Sommer as "Exposition of a New Theory on the Measurement of Risk," *Econometrica* 22 (1954): 22–36.

23. The St. Petersburg game continues to be provocative, and there are many perspectives on it. For more, we suggest you look at R. J. Aumann, "The St. Petersburg Paradox: A Discussion of Some Recent Comments," *Journal of Economic Theory* 14 (1977): 443–45, and P. Samuelson, "The St. Petersburg Paradox: Defanged, Dissected and Historically Described," *Journal of Economic Literature* 15 (1977): 24–55.

24. J. von Neumann and O. Morgenstern, *Theory of Games and Economic Behavior* (Princeton: Princeton University Press, 1944).

25. F. J. Anscombe and R. J. Aumann, "A Definition of Subjective Probability," *Annals of Mathematical Statistics* 34 (1964): 199–205.

26. F. P. Ramsey, "Truth and Probability" in *The Foundations of Mathematics and Other Logical Essays* (London: Routledge and Kegan Paul, 1931): 156–98.

27. L. J. Savage, *The Foundations of Statistics* (New York: Wiley, 1954).

28. See Richard Jeffrey and Michael Hendrickson, "Probabilizing Pathology," *Proceedings of the Aristotelian Society* (1989): 211–26. DOI: http://dx.doi.org/10.1093/aristotelian/89.1.211.

29. B. Skyrms, "Dynamic Coherence and Probability Kinematics," *Philosophy of Science* 54 (1987): 1–20. Reprinted in B. Skyrms, *From Zeno to Arbitrage* (New York and London: Oxford University Press, 2012).

CHAPTER 3

1. We cannot help but remark that people have systematic problems with deduction as well. Some deviations from correct logic are common enough to have become named fallacies. There are problems with conditionals and quantifiers. Students often become confused when dealing with Aristotelian syllogisms. For many examples, and one psychological theory designed to explain them, see P. Johnson-Laird and R. Byrne, *Deduction* (Mahwah, NJ: Erlbaum, 1991).

2. D. Laibson and R. Zeckhauser, "Amos Tversky and the Ascent of Behavioral Economics," *Journal of Risk and Uncertainty* 16 (1998): 7–47.

3. See, for instance, J. Liebman and R. Zeckhauser, "Simple Humans, Complex Insurance, Simple Subsidies," NBER working paper 14330 (Cambridge, MA: National Bureau of Economic Research, 2008).

4. See M. Allais, "Le comportement de l'homme rationnel devant le risque: critique des postulats et axiomes de l'école Américaine," *Econometrica* 21 no. 4 (1953): 503–46.

5. Paul Samuelson and Kenneth Arrow, for instance, made choices consistent with expected utility theory.

6. D. Ellsberg, "Risk, Ambiguity and the Savage Axioms," *Quarterly Journal of Economics* 75 (1961): 643–69.

7. If you don't know what the Pentagon papers are, you should find out.

8. J. M. Keynes, *A Treatise on Probability* (London: Macmillan, 1921).

9. F. H. Knight, *Risk Uncertainty and Profit* (Boston: Houghton Mifflin, 1921).

10. J. S. Mill, *A System of Logic: Ratiocinative and Inductive* (London: Harrison, 1843).

11. A. Tversky and D. Kahneman, "Judgment under Uncertainty: Heuristics and Biases," *Science* 185 (1974): 1124–31.

12. D. Kahneman, *Thinking Fast and Slow* (New York: Farrar, Strauss and Giroux, 2011). The two-process view that animates this enjoyable book should be seen as a deliberate oversimplification, and we recommend that the interested reader also consult the original papers.

13. D. Kahneman and A. Tversky, "Choices, Values and Frames," *American Psychologist* 39 (1984): 341–50.

14. B. J. McNeil, S. J. Pauker, H. C. Sox Jr., and A. Tversky, "On the Elicitation of Preferences for Alternative Therapies," *New England Journal of Medicine* 306 (1982): 1259–62.

15. H. C. Sox, M. C. Higgens, and D. K. Owen, *Medical Decision Making* (New York: Wiley, 2013).

16. Amos Tversky and Daniel Kahneman, "Rational Choice and the Framing of Decisions," *Journal of Business* 59 (1986): S251–78.

17. *The Port Royal Logic*, tr. Thomas Spencer Baynes (Edinburgh: James Gordon, 1861): 367–68.

18. H. Raiffa, *Decision Analysis* (Reading, MA: Addison Wesley, 1968).

CHAPTER 4

1. See E. D. Sylla, "The Emergence of Probability from the Perspective of the Leibniz–Jacob Bernoulli Correspondence," *Perspectives on Science* 6.1 and 2 (1998): 41–76.

2. Written about 1689, published posthumously in 1713. English translation by E. D. Sylla as J. Bernoulli, *The Art of Conjecturing, Together with Letter to a Friend on Sets in Court Tennis* (Baltimore: Johns Hopkins University Press, 2006).

3. Stigler suggests that this may be why Bernoulli did not publish *Ars Conjectandi* in his lifetime. See S. Stigler, *The History of Statistics: The Measurement of Uncertainty before 1900* (Cambridge, MA: Harvard University Press, 1990).

4. A. A. Cournot, *Exposition de la théory des chances et des probabilités* (Paris: Hachett, 1843). See the discussion of the history of this idea in G. Shafer and V. Vovk, "The Sources of Kolmogorov's *Grundbegriffe*," *Statistical Science* 21 (2006): 70–98.

5. Thanks to Steve Stigler for pointing out the transition to us.

6. The Hershel connection is pointed out in J. V. Strong, "John Stuart Mill, John Herschel, and the Probability of Causes," *PSA: Proceedings of the Biennial Meeting of the Philosophy of Science Association*, V. 1 (1978): 31–41. Strong refers to a letter from Herschel to Mill in December 1845. There are also two further letters supporting Laplace in April 1846. Mill replies that the second letter has convinced him. The letters are in the archives of the Royal Society.

7. J. Venn, *The Logic of Chance: An Essay on the Foundations and Province of the Theory of Probability with Especial Reference to its Application in Moral and Social Sciences* (London and Cambridge: Macmillan, 1866).

8. Maurice Fréchet, "The Diverse Conceptions of Probability," *Erkenntnis* 1 (1939): 7–23.

CHAPTER 5

1. A. N. Kolmogorov (1933), *Grundbegriffe der Wahrscheinlichkeitsrechnung*, Springer. English translation as *Foundations of the Theory of Probability* (New York: Chelsea, 1950).

2. A wonderful book full of such arguments is Mark Levi, *The Mathematical Mechanic* (Princeton: Princeton University Press, 2009). Some of the arguments are quite compelling, but they are different sorts of arguments than mathematical proofs.

3. M. Kac, *Enigmas of Chance* (New York: Harper and Row, 1985).

4. Countably additive. Kolmogorov's terminology is "completely additive."

5. The postulate of countable additivity arrives in two stages. First, finite additivity is postulated as all that is needed for finite spaces. Then Kolmogorov introduces a postulate of *continuity*:

If we have a decreasing series of sets whose intersection is empty, the limit of the probabilities of these sets as they become smaller must be zero.

He then shows that in the presence of the other axioms, this axiom is equivalent to countable additivity.

6. The existence of such conditional probabilities is guaranteed by the abstract Radon-Nikodym theorem, proved by Otto Nikodym in 1930.

7. The measurable space need not use [0, 1) as in our example, but some assumptions on the space are necessary.

8. For instance, J. L Doob, *Stochastic Processes* (New York: Wiley, 1953). See also Doob's memoire, "Kolmogorov's Early Work on Convergence Theory and Foundations," *Annals of Probability* 17 (1989): 815–21.

9. New York: Random House.

10. Princeton: Princeton University Press.

11. D. Freedman, *Statistical Models: Theory and Practice* (New York: Cambridge University Press, 2005) and *Statistical Models and Causal Inference: A Dialogue with the Social Sciences* (New York: Cambridge University Press, 2009).

12. See the discussion in S. L. Zabell, *Symmetry and its Discontents: Essays in the History of Inductive Probability* (New York: Cambridge University Press, 2005).

13. A. Berger and T. P. Hill, *An Introduction to Benford's Law* (Princeton: Princeton University Press, 2015).

14. Things are all right if *A* and *B* are disjoint; density is finitely additive. But it is possible that both *A* and *B* have a density, but their union does not. See R. C. Buck, "The Measure-Theoretic Approach to Density," *American Journal of Mathematics* 68 (1946): 560–80.

15. But it is worth mentioning that in the limit, as *s* tends to 1, there are sets for which the limit doesn't exist.

16. A. N. Kolmogorov, "Algèbres de Boole métriques complètes," *Zjazd Matematyków Polskich*. Appendix to the Annals of the Polish Society of Mathematicians 20 (1948): 21–30. Translated by R. Jeffrey as "Complete Metric Boolean Algebras," *Philosophical Studies* 77 (1995): 57–66.

17. A. N. Kolmogorov, "On Tables of Random Numbers," *Sankhya* 25 (1963): 369–75.

18. Fourth century BCE.

19. G. Vitali, *Sul problema della misura dei gruppi di punti di una retta* (Bologna: Gamberini e Parneggiani, 1965).

20. This is called "translation invariance."

21. Nonmeasurable in the senses of Borel and of Lebesgue.

22. F. Hausdorff, *Grundzüge der Mengenlehre* (Leipzig: Veit., 1914).

23. S. Banach and A. Tarski, "Sur la décomposition des ensembles de points en parties respectivement congruentes," *Fundamenta Mathematicae* 6 (1924): 244–77.

24. S. Banach and C. Kuratowski, "Sur une généralisation du problème de la mesure," *Fundamenta Mathematicae* 5 no. 14 (1929): 127ff.

CHAPTER 6

1. T. Bayes, "An Essay towards solving a Problem in the Doctrine of Chances," *Philosophical Transactions of the Royal Society* 53 (1763): 370–418.

2. S. Stigler, "True Title of Bayes' Essay," *Statistical Science* 28 (2013): 283–88.

3. Quoted in Raynor, "Hume's Knowledge of Bayes's Theorem," *Philosophical Studies* 38 (1980): 105–6.

4. In statistical jargon, the chance of seeing j successes out of n, $\binom{n}{j} x^j (1-x)^{n-j}$ is called the likelihood. One natural choice of prior has the same form with j and $n-j$ now replaced by two parameters, $\alpha - 1$ and $\beta - 1$, with α and β positive numbers. This is now thought of as a function of x and normalized so that it integrates to 1. This is called a β prior (with parameters α and β).

5. P. Diaconis and D. Ylvisaker, "Quantifying Prior Opinion," in J. M. Bernardo, M. H. Degroot, D. V. Lindley, A. F. M. Smith, eds., *Bayesian Statistics 2: Proceedings of the Second Valencia International Meeting* (North-Holland-Amsterdam: Elsevier, 1985): 133–56.

6. P. Diaconis and D. Freedman, "On the Consistency of Bayes Estimates," *Annals of Statistics* 14 (1986): 1–26.

7. Open Science Collaboration, "Estimating the Reproducibility of Psychological Science," *Science* 28 Aug. 2015, Vol. 349 Issue 6251. DOI: 10.1126/science.aac4716.

8. J.P.A. Ioannidis, "Why Most Published Research Findings Are False," *PLoS Medicine* 2 no. 8 (2005): e124.

9. C. Glenn Begley and Lee M. Ellis, "Drug Development: Raise Standards for Preclinical Research," *Nature* 483 (2012): 531–33. DOI: 10.1038/483531a.

10. Ioannidis (2005), 0696.

11. H. Pashler and C. Harris, "Is the Replicability Crisis Overblown? Three Arguments Examined," *Perspectives on Psychological Science* 7 (2012): 531–36. DOI: 10.1177/1745691612463401.

12. There is an interactive p-hacking app on Nate Silver's blog that you might like to try: http://fivethirtyeight.com/features/science-isnt-broken/#part2.

13. See S. L. Zabell, "Commentary on Alan M. Turing: The Applications of Probability to Cryptography," *Cryptologia* 36 no. 3 (2012): 191–214.

14. B. de Finetti (New York: Wiley, 1972).

15. For a careful treatment of how Bayesian thinking affects even such basic probability problems, see P. Diaconis and S. Holmes, "A Bayesian Peek into Feller, Volume I," *Sankhya* 64 (2002): 820–41.

16. But if you have information that makes for a nonuniform prior, why leave it out?

17. Technically, minimax or admissible.

18. J. Cornfield, "The Bayesian Outlook and its Applications," *Biometrics* 25 (1969): 617–57.

19. Details can be found in Jim Berger's *Statistical Decision Theory* (New York: Springer Berger, 1980), and Wolpert's *The Likelihood Principle* (Bethesda, MD: Institute of Mathematical Statistics, 1988) [free online at Project Euclid of the IMS] is also highly recommended.

20. For the amazing story of Dorothy Wrinch, see D. Senechal, *I Died for Beauty: Dorothy Wrinch and the Cultures of Science* (New York: Oxford University Press, 2012). For the joint work of Jeffreys and Wrinch: "On Some Aspects of the Theory of Probability," *Philosophical Magazine* 38 (1919): 715–31; "On Certain Fundamental Principles of Scientific Inquiry," *Philosophical Magazine* 42 (1921): 369–90; "On Certain Fundamental Principles of Scientific Inquiry," *Philosophical Magazine* 45 (1923): 368–74.

21. How does one ever learn a true universal generalization? This is a criticism that was made of the (1950) system of inductive logic of the philosopher Rudolf Carnap, which is essentially rolling a die with a uniform prior. For discussion see S. L. Zabell, "Confirming Universal Generalizations," *Erkenntnis* 45 (1996): 267–83.

22. The example is in H. Jeffreys, *Theory of Probability*, 2d ed. (1949), 3d ed. (1961), (Oxford: Clarendon Press, 1939).

23. I. J. Good, *Probability and the Weighing of Evidence* (London: Griffin, 1950).

24. Statisticians do many things other than estimating parameters (the i's from before). They design experiments, data displays, and ways with dealing with miss-specified models, wildly outlying data, and missing data. For pointers to philosophical aspects see D. Freedman, *Statistical Models: Theory and Practice* (New York: Cambridge University Press, 2005); P. Diaconis, "Theories of Data Analysis: from Magical Thinking through Classical Statistics," in D. C. Hoaglin

et al., eds, *Exploring Data Tables Trends and Shapes* (New York: Wiley, 1985); and P. Diaconis, "A Place for Philosophy? The Rise of Modeling in Statistics," *Quarterly Journal of Applied Mathematics* 56 (1998): 797–805.

25. D. Cox, *Principles of Statistical Inference* (New York: Cambridge University Press, 2006).

26. See P. Diaconis, "Theories of Data Analysis, From Magical Thinking to Mathematical Statistics," in D. C. Hoaglin et al., ed., *Exploring Data Tables Trends and Shapes* (New York: Wiley, 1985).

CHAPTER 7

1. New York: Wiley, 1972.

2. "La Prévision: ses lois logiques, ses sources subjectives," *Annales de l'Institut Henri Poincaré* (1937), tr. as "Foresight: Its Logical Laws, Its Subjective Sources," in H. E. Kyburg and H. E. Smokler, eds., *Studies in Subjective Probability* (New York: Wiley): 196.

3. Bruno de Finetti, "Sur la condition d'équivalence partielle," *Actualités scientifiques et Industrielles* 739 (Hermann & Cie, 1938). Translated as "On the Condition of Partial Exchangeability" by P. Benacerraf and R. Jeffrey in *Studies in Inductive Logic and Probability II*, ed. R. Jeffrey (Berkeley: University of California Press, 1980): 193–205.

4. A detailed commentary in modern language is in S. Bacallando, P. Diaconis, and S. Holmes, "De Finetti's Priors Using Markov Chain Monte Carlo Computations," *Statistics and Computing* 25, no. 4 (2015): 797–808.

5. See P. Diaconis, "Finite Forms of de Finetti's Theorem," *Synthese* (1977): 271–81.

6. It is called nonparametric Bayesian statistics. See J. K. Ghosh and R. V. Ramamoorthy, *Bayesian Non-Parametrics* (Berlin: Springer, 2003) for a textbook account.

7. For more, see P. Diaconis and D. Freedman, "Partial Exchangeability and Sufficiency," in *Statistics: Applications and New Directions* (Calcutta: Indian Statistical Institute, 1981), and O. Kallenberg, *Probabilistic Symmetries and Invariance Principles* (Berlin: Springer, 2005).

8. D. Freedman, "Mixtures of Markov Processes," *Annals of Mathematical Statistics* 2 (1962): 615–29.

9. P. Diaconis and D. Freedman, "de Finetti's Theorem for Markov Chains," *Annals of Probability* 8 (1980): 115–30.

10. B. de Finetti, "Sur la condition d'équivalence partielle," *Actualités scientifiques et Industrielles* 739 (Hermann & Cie, 1938). Translated as "On the Condition of Partial Exchangeability" by P. Benacerraf and R. Jeffrey in *Studies in Inductive Logic and Probability II*, ed. R. Jeffrey (Berkeley: University of California Press, 1980): 193–205.

11. G. Birkhoff, "Proof of the Ergodic Theorem," *Proceedings of the National Academy of Sciences of the USA* 17 (1931): 656–60.

12. H. Weyl, *Symmetry* (Princeton: Princeton University Press, 1952).

13. See D. Freedman, "Invariants under Mixing which Generalize de Finetti's Theorem," *Annals of Mathematical Statistics* 33 (1962): 916–23.

CHAPTER 8

1. G. Marsaglia, "Random Numbers Fall Mainly in the Planes," *Proceedings of the National Academy of Sciences of the USA* 61 (1968): 25–28.

2. D. Knuth, *The Art of Computer Programming v. III* (Reading, MA: Addison Wesley, 1973).

3. M. Matsumoto and T. Nishimura, "Mersenne Twister: a 623-dimensionally Equidistributed Uniform Pseudo-random Number Generator," *ACM Transactions on Modeling and Computer Simulation* 8 (1998): 3–30.

4. "Coin Tossing Computers Found to Show Subtle Bias," *New York Times*, Jan. 12, 1993.

5. W. Press, S. A. Terkolsky, W. T. Vetterling, and B. P. Flannery, *Numerical Recipes: The Art of Scientific Computing*, 3rd ed. (Cambridge: Cambridge University Press, 1987).

6. M. Blum and S. Micali, "How to Generate Cryptographically Strong Sequences of Pseudorandom Bits," *SIAM Journal on Computing* 13 (1984): 850–64.

7. P. Martin-Löf, "The Definition of Random Sequences," *Information and Control* 9 (1966): 602–19.

8. Notably Fréchet.

9. A. Wald, "Die Widerspruchsfreiheit des Kollektiv begriffes der Wahrscheinlichkeitsrechnung," *Ergebniss eines mathematischen Kolloquiums* 8 (1937): 38–72.

10. A. Church, "On the Concept of a Random Sequence," *Bulletin of the American Mathematical Society* 40 (1940): 254–60.

11. R. von Mises, *Probability, Statistics and Truth* (London: Macmillan, 1939).

12. J. Ville, *Étude critique de la notion de collectif*, Monographies des Probabilités 3 (Paris: Gauthier-Villars, 1939).

13. P. Martin-Löf, "The Definition of Random Sequences," *Information and Control* 9 (1966): 602–19.

14. D. Hilbert and W. Ackermann, *Grundzüge der theoretischen Logik* (Berlin: Springer, 1928).

15. A. Church, "An Unsolvable Problem of Elementary Number Theory," *American Journal of Mathematics* 58 (1936): 345–63. A. Turing, "On Computable Numbers, with an Application to the Entscheidungsproblem," *Proceedings of the London Mathematical Society* Series 2, 42 (1937): 230–65.

16. A. A. Markov, *The Theory of Algorithms*, American Mathematical Society Translations, series 2, 15 (1960): 1–14.

17. K. Gödel (1946), "Remarks before the Princeton Bicentenial Conference on Problems in Mathematics," in *Collected Works v. II*, ed. S. Feferman (New York: Oxford University Press, 1990): 150.

18. In fact, Martin-Löf shows that there is a universal test such that if a sequence passes it, it passes all tests.

19. C. P. Schnorr, *Zufälligkeit und Wahrscheinlechkeit: Eine algorithmische Begründung der Wahrscheinlichkeitstheorie* (Berlin: Springer, 1971).

20. A. N. Kolmogorov, "On Tables of Random Numbers," *Sankhyā, The Indian Journal of Statistics Ser. A* 25 (1963): 369–76.

21. G. J. Chaitin, "On the Length of Programs for Computing Finite Binary Sequences: Statistical Considerations," *Journal of the Association for Computing Machinery* 16 (1966): 407–22.

22. R. J. Solomonoff, "A Preliminary Report on a General Theory of Inductive Inference," Technical Report ZTB-138 (Cambridge, MA: Zator, 1960) and "A Formal Theory of Inductive Inference Parts I and II," *Information and Control* 7 (1964): 1–22 and 224–54.

23. R. J. Solomonoff, "Algorithmic Probability—Its Discovery—Its Properties and Application to Strong AI," chapter 11 in Hector Zenil, ed., *Randomness through Computation* (Singapore: World Scientific, 2011): 149–57.

24. L. Levin, "On the Notion of a Random Sequence," *Soviet Mathematics Doklady* 14 (1973): 1413–16.

25. G. J. Chaitin, "A Theory of Program Size Formally Identical to Information Theory," *Journal of the Association for Computing Machinery* 22 (1975): 329–40.

CHAPTER 9

1. L. Boltzmann, "Weitere Studien über das Wärmegleichgewicht unter Gasmolekün," *Sitzungberichte der Akademie der Wissenschaften zu Wien, mathematischnaturwissenschaftliche Klasse* 66 (1872): 275–370. Tr. in S. G. Brush, *The Kinetic Theory of Gases* (London: Imperial College Press, 2003).

2. J. Loschmidt, "Über den Zustand des Wärmgleichgewichtes eines Systemes von Körpern mit Rücksicht auf die Schwerkraft," *Sitzungberichte der Akademie der Wissenschaften zu Wien, mathematisch-naturwissenschaftliche Klasse* 73 (1876): 128–42.

3. E. Zermelo, "Uber enien Satz der Dynamik und die mechanische Wärmetheorie," *Annalen der Physik* 57 (1896): 485–94. English translation in Brush (2003): 382–91.

4. H. Poincaré, "Sur le problème des trois corps et les équations de la dynamique," *Acta Mathematica* 13 (1890): 1–270. Partial English translation in Brush (2003): 368–76.

5. P. Ehrenfest and T. Ehrenfest-Afanassjewa, *The Conceptual Foundations of the Statistical Approach in Mechanics* (Ithaca: Cornell University Press, 1959); translation of "Begriffliche Grundlagen der statistischen Auffassung in der Mechanik," *Encyklopädie der mathematischen Wissenschaften*, Volume IV/2/II/6 (Leipzig: B. G. Teubner, 1912).

6. M. Kac, "Random Walk and the Theory of Brownian Motion," *American Mathematical Monthly* 54 (1947): 369–91.

7. This is Jaynes' "maximum entropy" approach applied to this toy problem. For Jaynes on statistical mechanics, see E. T. Jaynes, "Information Theory and Statistical Mechanics," *Physical Review* 106 no. 4 (1967): 620–30, and "Information Theory and Statistical Mechanics II," *Physical Review* 108 no. 2: 171–90.

8. See P. Diaconis and M. Shahshahani, "Time to Reach Sationarity in the Bernoulli-Laplace Diffusion Model," *Siam Journal of Mathematical Analysis* 18 (1987): 202–18.

9. Or arbitrarily close to every point in phase space.

10. Iterating the transformation gives a semigroup of transformations if an initial time is presupposed or a group if time is infinite in past and future.

11. More precisely, that is to say that the symmetric difference between s and $T^1(s)$ has measure zero.

12. Lebesgue measure on the phase space restricted to a constant-energy hypersurface.

13. Liouville's theorem.

14. Ya. G. Sinai, "On the Foundation of the Ergodic Hypothesis for a Dynamical System of Statistical Mechanics," *Soviet Mathematics Doklady* 4 (1963): 1818–22.

15. Ya. G. Sinai and N. I. Chernov, "Ergodic Properties of Certain Systems of 2-D Discs and 3-D Balls," *Russian Mathematical Surveys* 42 (1987): 181–207.

16. English translation by A. R. Fuller (London: Taylor and Francis, 1992).

17. See E. Engel, *A Road to Randomness in Physical Systems*, Lecture Notes in Statistics 71 (Berlin: Springer, 1992).

18. Some dynamical systems can be proved to be mixing. E. Hopf used such proofs to justify the usual assumptions of independent repeated trials. See Engle, cited in the previous note, for an account.

19. For more on these topics, see D. Ornstein, *Ergodic Theory, Randomness and Dynamical Systems*, Yale Mathematical Monographs 5 (New Haven: Yale University Press, 1974) and D. Ornstein and B. Weiss, "Statistical Properties of Chaotic Systems," *Bulletin of the American Mathematical Society* 24 (1991): 11–116.

20. D. Szátz, "Boltzmann's Ergodic Hypothesis: A Conjecture for Centuries?" Mathematical Institute of the Hungarian Academy of Sciences (Budapest, Hungary, 1994).

21. For an invitation to think subjectively about quantum mechanics from a mainstream physicist, see D. Mermin, "Quantum Mechanics: Fixing the Shifty Split," 65 (2012): 12 (http://dx.doi.org/10.1063/PT.3.1618), which contains further references.

22. The interpretation of quantum mechanics has generated a huge philosophical literature, which has little to do with its applications. Theories that are metaphysically quite different aim at producing the very same probabilities of outcome conditional on measurement performed. As Mermin says, "New interpretations appear every year; none ever seems to disappear." (See note 21.)

23. Two good introductory accounts of quantum mechanics are L. Susskind and A. Friedman, *Quantum Mechanics: The Theoretical Minimum* (New York: Basic Books, 2014), and D. Griffiths, *Introduction to Quantum Mechanics*, 2d ed. (New York: Prentice Hall, 2004). For a friendly account with pointers to the philosophical literature see R.I.G. Hughes, *The Structure and Interpretation of Quantum Mechanics* (Cambridge, MA: Harvard University Press, 1992).

24. A. Einstein, B. Podolsky, and N. Rosen, "Can Quantum-Mechanical Description of Physical Reality Be Considered Complete?" *Physical Review* 47 (1935): 777–80.

25. J. S. Bell, "On the Einstein–Podolsky–Rosen Paradox," *Physics* 1 (1964): 195–200. Reprinted in J. S. Bell, *Speakable and Unspeakable in Quantum Mechanics* (Cambridge: Cambridge University Press, 2004): 14–21.

26. A singlet state.

27. Starting with A. Aspect, J. Dalibard, and G. Roger, "Experimental Test of Bell's Inequalities Using Time-Varying Analyzers," *Physical Review Letters* 49 (1982): 1804–7.

28. W. Tittel, J. Brendel, H. Zbinden, and N. Gisin, "Violation of Bell Inequalities by Photons More Than 10 km Apart," *Physical Review Letters* 81 (1998): 3563–66.

29. G. Weihs, T. Jennewein, C. Simon, H. Weinfurter, and A. Zeilinger, "Violation of Bell's Inequality under Strict Einstein Locality Conditions," *Physical Review Letters* (1998): 5039–43.

30. For discussion of how the experiments close loopholes, see the Scholarpedia article on Bell's theorem, by S. Goldstein, T. Norsen, D. V. Tausk, and N. Zanghi, doi:10.4249/scholarpedia.8378, and the Stanford Encyclopedia of Philosophy articles by A. Shimony and by A. Fine.

31. This is not to say that it is impossible to have a deterministic hidden variable theory that recovers the quantum probabilities. Such theories exist. But they do not recover the locality that was Einstein, Podolsky, and Rosen's original motivation for considering them. See D. Bohm, "A Suggested Interpretation of the Quantum Theory in Terms of 'Hidden' Variables, I and II," *Physical Review* 85 (1952): 166–79, 180–93. In Bohmian mechanics, the underlying physical process is deterministic, and the quantum mechanical probabilites are purely epistemic. We will return to this in the appendix.

32. The study of how quantum mechanical systems behave in an environment in which their classical counterparts are chaotic was dubbed quantum chaology in M. V. Berry, "The Bakerian Lecture 1987: Quantum Chaology," *Proceedings of the Royal Society of London. Series A, Mathematical and Physical Sciences* 413 (1987): 183–98.

33. L. A. Bunimovich, "On the Ergodic Properties of Nowhere Dispersing Billiards," *Communications Mathematical Physics* 65 (1979): 295–312.

34. To even state it correctly requires more machinery than we have in this book. For a review article, see S. Nonnenmacher, "Anatomy of Quantum Chaotic Eigenstates," *Seminaire Poincare* XIV (2010): 177–220.

35. Thus, the stadium, being ergodic, is quantum ergodic. But the ghost of the exceptional "bouncing ball" orbits remains in the "scarring" visible in some eigenstates. That this scarring is genuine is proved by A. Hassell. See Hassell's introduction to the subject, "What is Quantum Unique Ergodicity?" *Australian Mathematical Society Technical Paper*.

36. See H. D. Zeh, "On the Interpretation of Measurement in Quantum Theory," *Foundations of Physics* 1 (1970): 69–76.

37. For a lucid discussion of these issues and advocacy of the Bohmian approach, see J. S. Bell, *Speakable and Unspeakable in Quantum Mechanics: Collected Papers* (Cambridge: Cambridge University Press, 2004).

38. D. Bohm, "A Suggested Interpretation of the Quantum Theory in Terms of 'Hidden' Variables, I and II," *Physical Review* 85 (1952): 166–93.

39. The reason being "decoherence"; the interaction destroys the coherence required for interference effects.

40. H. Everett III, " 'Relative State' Formulation of Quantum Mechanics," *Review of Modern Physics* 29 (1957): 454–62. For Everett's recently discovered nachlass, see J. A. Barrett and P. Byrne, eds., *The Everett Interpretation of Quantum Mechanics: Collected Works 1955–1980 with Commentary* (Princeton: Princeton University Press, 2012). Everett's manuscripts are archived at the University of California, Irvine Libraries, available online at http://hdl.handle.net/10575/1060.

41. For further discussion see J. A. Barrett, *The Quantum Mechanics of Minds and Worlds* (New York: Oxford University Press, 1999); S. Saunders, J. Barrett, A. Kent, and D. Wallace, *Many Worlds: Everett, Quantum Theory and Reality* (New York: Oxford University Press, 2012); D. Wallace, *The Emergent Multiverse: Quantum Theory According to the Everett Interpretation* (Oxford: Oxford University Press, 2014).

42. Wallace (previous footnote) uses a rational degree-of-belief interpretation; Everett is a kind of frequentist.

CHAPTER 10

1. B. Russell, *The Problems of Philosophy*, Home University Library (London: Williams and Norgate, 1912). Available online at Project Gutenberg.

2. L. J. Savage, "Implications of Personal Probability for Induction," *Journal of Philosophy* 64 (1967): 593–607.

3. B. de Finetti, "Sur la condition d'équivalence partielle," *Actualités scientifiques et Industrielles* 739 (Hermann & Cie, 1938). Translated as "On the Condition of Partial Exchangeability," by P. Benacerraf and R. Jeffrey in *Studies in Inductive Logic and Probability II*, ed. R. Jeffrey (Berkeley: University of California Press, 1980): 193–205.

4. D. Hume, *An Enquiry Concerning Human Understanding* (London: Millar, 1748/1777), Section IV, Part I, 21.

5. Hume (1748/1777) IV, Part II, 32.

6. For a nice entrée into the literature, with a guide to diverging interpretations, see G. De Pierris and M. Friedman, "Kant and Hume on Causality," *The Stanford Encyclopedia of Philosophy*, https://plato.stanford.edu/entries/kant-humecausality/.

7. C. D. Broad, *The Philosophy of Francis Bacon* (Cambridge: Cambridge University Press, 1926), http://www.ditext.com/broad/bacon.html.

8. K. Popper, *Logik der Forshung* (Vienna: Springer Verlag). Translated as *The Logic of Scientific Discovery* (London: Hutchinson) and *The Logic of Scientific Discovery*, 2d. ed. (New York: Harper, 1934/1959/1968).

9. Popper (1968): 29.

10. First and second centuries AD. Most of what we know of him is through the account of Diogenes Laërtius, *Lives of the Eminent Philosophers*, Book IX 90, tr. R. D. Hicks, Loeb Classical Library (Cambridge, MA: Harvard University Press, 1925).

11. See B. Skyrms, "Grades of Inductive Skepticism," *Philosophy of Science* 81 (2014): 303–12, from which this chapter is largely drawn.

12. Bernoulli (1713/2005): chapter 4.

13. De Moivre made similar claims in the second and third editions of *The Doctrine of Chances*.

14. R. Price, Preface to "An Essay Towards Solving a Problem in the Doctrine of Chances," *Proceedings of the Royal Society of London* 53 (1763): 370–76.

15. Hume (1749/1777), Section VI, 47.

16. See D. A. Gillies, "Was Bayes a Bayesian?" *Historia Mathematica* 14 (1987): 325–46, S. L. Zabell, "The Rule of Succession," *Erkenntnis* 31 (1989): 283–321, and S. L. Zabell, "The Continuum of Inductive Methods Revisited," in *The Cosmos of Science*, ed. J. Norton and J. Earman (Pittsburgh: University of Pittsburgh Press, 1997): 351–85.

17. S. Stigler, "True Title of Bayes' Essay," *Statistical Science* 28 (2013): 283–88.

18. P. S. Laplace, "Mémoire sur la probabilité des causes par les événemens," *Savants étranges* 6 (1774): 621–56. Tr. by Stephen Stigler as "Memoir on the Probability of the Causes of Events," *Statistical Science* 1 (1986): 359–78.

19. As were Kurt Grelling and Carl Hempel.

20. H. Reichenbach, *Experience and Prediction* (Chicago: University of Chicago Press, 1938): 352.

21. In fact, he proves something stronger, the Bernstein–von Mises phenomenon. The posterior is asymptotically normal.

22. Laplace (1774/1986).

23. Here we mean the skeptics who succeeded Plato in the Academy, starting with Arcesileas and Carneades. There is a line of influence through Cicero and to modern times that makes a fascinating story. As an entry point, we recommend R. Popkin's *History of Skepticism*, rev. ed. (New York: Oxford University Press, 2003).

24. In the weak-star topology. This is defined in terms of convergence of expectations, which are what we usually care about. A sequence of probability measures, P_n, converges weak-star to P if, for all bounded continuous functions f, the associated expectations converge: $E_n(f) \longrightarrow E(f)$.

25. This is not to say that these results hold with absolute generality. With an infinite number of categories, things are more complicated. There are still consistent "ignorance" priors, but their characterization is not so straightforward or intuitively compelling. See P. Diaconis and D. Freedman, "On the Consistency of Bayes' Estimates," *Annals of Statistics* 14 (1988): 1–26.

26. J. L. Doob, "Application of the Theory of Martingales," *Actes du Colloque International Le Calcul des Probabilités et ses applications* (Paris CNRS, 1948): 23–27.

27. Hume (1749/1777), Section VI, 46.

28. De Finetti's theorem has been proved in considerably more general form. See E. Hewitt and L. J. Savage, "Symmetric Measures on Cartesian Products," *Transactions of the American Mathematical Society* 80 (1955): 470–501.

29. You may be skeptical about infinite sequences being part of the world, as was de Finetti. If so, consider finite exchangeable sequences that can be extended to longer sequences that remain exchangeable. See P. Diaconis and D. Freedman, "Finite Exchangeable Sequences," *The Annals of Probability* 8 (1980): 745–64. On de Finetti's finitist views of his theorem, see S. L. Zabell, "De Finetti, Chance and Quantum Physics," in *Bruno de Finetti: Radical Probabilist*, ed. Maria Carla Galavotti (London: College Publications, 2009): 59–83, and D. M. Cifarelli and E. Regazzini, "De Finetti's Contribution to Probability and Statistics," *Statistical Science* 11 (1996): 2253–82.

30. B. de Finetti, "Sur la condition d'équivalence partielle," *Actualités Scientifiques et Industrialles* 739 (Hermann & Cie, 1938), tr. Paul Benacerraf and Richard Jeffrey as "On the Condition of Partial Exchangeability," in *Studies in Inductive Logic and Probability II*, ed. Richard Jeffrey (Berkeley: University of California Press, 1980): 193–205.

31. See P. Diaconis and D. Freedman, "De Finetti's Generalizations of Exchangeability," in *Studies in Inductive Logic and Probability II*.

32. As suggested in P. Billingsley, *Ergodic Theory and Information* (New York: Wiley, 1965).

33. More generally, of the average values of a measurable function.

34. That is, for almost every point in the space—with "almost every" determined by your probabilities—the limiting relative frequencies exist.

35. The limiting relative frequency of *A* is a random variable. Your expectation of this limiting relative frequency is your probability of A. In the special case in which your degrees of belief are ergodic, *you are sure that your probability of A is equal to the limiting relative frequency.* This is Birkhoff's ergodic theorem.

36. N. Goodman, *Fact, Fiction and Forecast* (Cambridge, MA: Harvard University Press, 1955).

37. See B. Skyrms, "Bayesian Projectibility," in *Grue! The New Riddle of Induction*, ed. D. Stalker (Chicago: Open Court, 1994). Reprinted in Skyrms, *From Zeno to Arbitrage* (Oxford: Oxford University Press, 2012), chapter 13, for a more thorough discussion.

38. A measurable set in one's probability space.

39. R. Jeffrey, *The Logic of Decision* (New York: McGraw-Hill, 1965) and "Probable Knowledge" in I. Lakatos, ed., *The Problem of Inductive Logic* (Amsterdam: North Holland, 1968).

40. M. Goldstein, "The Prevision of a Prevision," *Journal of the American Statistical Association* 78 (1983): 817–19.

41. B. van Fraassen, "Belief and the Will" *The Journal of Philosophy* 81 (1984): 235–56.

42. B. Skyrms, *The Dynamics of Rational Deliberation* (Cambridge, MA: Harvard University Press, 1990) and "Diachronic Coherence and Radical Probabilism," *Philosophy of Science* 73 (2006): 959–68, reprinted in Skyrms, *From Zeno to Arbitrage* (New York: Oxford University Press, 2012).

43. S. L. Zabell, "It All Adds Up: The Dynamic Coherence of Radical Probabilism," *Philosophy of Science* 69 (2002): S98–S103.

ANNOTATED SELECT BIBLIOGRAPHY

CHAPTER 1

Bernoulli, J. *Ars Conjectandi*. Tr. Edith Dudley Sylla as *The Art of Conjecturing, Together with a Letter to a Friend on Sets in Court Tennis*. Baltimore: Johns Hopkins University Press, 2006.
> Translated with extensive scholarly notes, this reprises Pascal, Fermat, and Huygens and then goes on to prove the weak law of large numbers. It is a great achievement, but you will also find the original of Bernoulli's swindle.

Feller, W. *An Introduction to Probability Theory and its Applications, Vol. I*. New York: Wiley, 1957.
> Modify the classic view to allow unequal faces of a die (as suggested by Newton) and you have tools to explore all the discrete probability in this classic.

Huygens, C. *On Reasoning in Games of Chance* (available online at https://math.dartmouth.edu/~doyle/docs/huygens/huygens.pdf).
> A great scientist writes the first textbook on probability theory, after learning the contributions of Pascal and Fermat.

Ore, O. *Cardano, the Gambling Scholar*. Princeton: Princeton University Press, 1953.

A fascinating account of Cardano's life, together with a translation of his book, which blends probability theory with practical advice against cheaters.

CHAPTER 2

Anscombe, F. J., and R. J. Aumann. "A Definition of Subjective Probability," *Annals of Mathematical Statistics* 34 (1964): 199–205.
If you presuppose the existence of objective chances, the treatment of subjective probability is easy and elegant.

de Finetti, B. "Foresight: Its Logical Laws, Its Subjective Sources," in *Annales de l'Institut Henri Poincaré* 7 (1937): 1–68. English translation by H. E. Kyburg Jr. and H. E. Smokler, eds., in *Studies in Subjective Probability*. Huntington, NY: Krieger, 1964; 2d ed. 1980.
Kyburg consulted with de Finetti on the translation. See de Finetti's footnote to Ramsey's essay here, which he did not know when he wrote the original. This has both the Dutch book and the de Finetti representation theorem.

Freedman, D., and R. Purves. "Bayes Method for Bookies," *Annals of Mathematical Statistics* 40 (1969): 1177–86.
But you don't have the priors! is a kneejerk criticism of Bayesian updating. Freedman and Purves show that any coherent alternative methodology must behave *as if* there are priors and updating is by conditioning.

Ramsey, F. P. "Truth and Probability," 1926. In Ramsey, *The Foundations of Mathematics and other Logical Essays*, Ch. VII, ed. R. B. Braithwaite. London: Kegan, Paul, Trench, Trubner; New York: Harcourt, Brace, 1931: 156–98.
This essay was a written presentation to a philosophy club at Cambridge. In a very clear and readable way, it introduces the ideas for a simultaneous representation for both subjective probability and subjective utility. The depth of this work was only appreciated decades later.

Savage, L. J. *The Foundations of Statistics.* New York: Wiley, 1954.
The modern classic establishing simultaneous foundations of subjective probability and subjective utility. It also discusses the application of foundations to practical statistics.

CHAPTER 3

Ellsberg, D. "Risk, Ambiguity, and the Savage Axioms," *Quarterly Journal of Economics* 75 (1961): 643–69.
> Read this and then the commentary by Howard Raiffa.

Laibson, D., and R. Zeckhauser. "Amos Tversky and the Ascent of Behavioral Economics," *Journal of Risk and Uncertainty* 16 (1998): 7–47.
> Hear the story from the horse's mouth.

Raiffa, H. "Risk, Ambiguity, and the Savage Axioms: Comment," *Quarterly Journal of Economics* 75 (1961): 690–94.
> Read this comment after Ellsberg (1961).

Sox, H. C., M. C. Higgens, and D. K. Owen. *Medical Decision Making.* New York: Wiley, 2013.
> Tversky and Kahneman showed some systematic biases in life-or-death medical decisions. This book, by sophisticated doctors, shows how systematic application of Bayesian methodology can improve medical practice.

Tversky, A. and D. Kahneman. "Judgment under Uncertainty: Heuristics and Biases" *Science* 185 (1974): 1124–31.
> An opening volley in the behavioral economics revolution (if it is a revolution). Try these on your friends.

CHAPTER 4

Church, A. "On the Concept of a Random Sequence," *Bulletin of the American Mathematical Society* 46 (1940): 254–60.
> In this short note Alonzo Church injects computability for the first time into the mathematical discussion of randomness.

Venn, J. *The Logic of Chance: An Essay on the Foundations and Province of the Theory of Probability with Especial Reference to its Application in Moral and Social Sciences*. London and Cambridge: Macmillan, 1966.

> This very readable book by the leading English frequentist of the nineteenth century starts out as a polemic against Laplacian subjectivism and a defense of probability as long-run relative frequency. But more and more difficulties crop up, and more and more modifications are required.

Von Mises, R. *Probability, Statistics and Truth*, 2d rev. English ed. New York: Macmillan, 1957.

> A nontechnical introduction to his philosophy of probability. This English translation of the 1928 original was edited by his widow Hilda Geiringer.

CHAPTER 5

Fréchet, M. "The Diverse Definitions of Probability," *The Journal of Unified Science (Erkenntnis)* 8 (1939): 7–23.

> This is a lecture by a major mathematical contributor to probability that was delivered to a philosophical congress in 1938. He discusses von Mises' frequency view, de Finetti's subjectivism, and a kind of propensity interpretation.

Kolmogorov, A. N. *Grundbegriffe der Wahrscheinlichkeitsrechnung*. Berlin: Springer, 1933. English translation as *Foundations of the Theory of Probability*. New York: Chelsea, 1950.

> If anything deserves to be called a classic, this does. It sets up the framework for modern probability theory. It starts gently, with the finite case, and then transitions to the measure theory for the general case.

CHAPTER 6

Bayes, T. "An Essay towards Solving a Problem in the Doctrine of Chances," *Philosophical Transactions of the Royal Society* 53 (1763): 370–418.

Bayes' essay is tough going for the reader used to modern mathematical exposition, but if you are willing to put in the effort, it is remarkable what you will find.

Johnson, W. E. "Probability: The Deductive and Inductive Problems," *Mind* 49 (1932): 409–23.

Before de Finetti, a Cambridge logician introduces the concept of exchangeability and uses it (together with another property) to derive a generalization of Laplace's rule of succession.

Laplace, P. S. "Mémoire sur la probabilité des causes par les événemens," *Savants étranges* 6 (1774): 621–56. Translated by Stephen Stigler as "Memoir on the Probability of the Causes of Events," *Statistical Science* 1 (1986): 359–78.

An altogether remarkable essay by the young Laplace.

CHAPTER 7

De Finetti, B. "Sur la condition d'équivalence partielle," *Actualités scientifiques et Industrielles* 739 (Hermann & Cie, 1938). Translated by P. Benacerraf and R. Jeffrey as "On the Condition of Partial Exchangeability," in *Studies in Inductive Logic and Probability* II, ed. R. Jeffrey. Berkeley: University of California Press, 1980: 193–205.

Already, in 1938, de Finetti sees the way to far reaching generalizations to his representation theorem.

Diaconis, P. "Finite Forms of de Finetti's Theorem on Exchangeability," *Synthese* 36 (1977): 271–81.

How close do we get to the ideal infinite case with only finite sequences of trials?

Diaconis. P., and D. Freedman. "De Finetti's Theorem for Markov Chains," *Annals of Probability* 8 (1980): 115–30.

Beyond independent trials, a de Finetti representation theorem for recurrent Markov-exchangeable sequences. This gives a rigorous treatment of ideas in de Finetti (1938).

CHAPTER 8

Bienvenu, L., G. Shafer, and A. Shen. "On the History of Martingales in the Study of Randomness," *Journ@l électronique d'Histoire des Probabilités et de la Statistique/Electronic Journal for History of Probability and Statistics* 5 (2009): 1–40.

> This is a very interesting essay, informed by personal interviews and new translations.

Downey, R., and D. Hirschfeldt. *Algorithmic Randomness and Complexity*. Berlin: Springer, 2010.

> The current "bible" on the subject, this is one place you can find an exposition of famous unpublished work on randomness that Solovay did in the 1970s.

Kolmogorov, A. N. "On Tables of Random Numbers," *Sankhyā, The Indian Journal of Statistics Ser. A* 25 (1963): 369–76.

> Kolmogorov introduces the computational complexity approach to randomness.

Marsaglia, G. "Random Numbers Fall Mainly in the Planes," *Proceedings of the National Academy of Sciences of the USA* 61 (1968): 25–28.

> This blew the whistle on what was at the time a very widely used random number generator.

Martin-Löf, P. "The Definition of Random Sequences," *Information and Control* 9 (1966): 602–19.

> Many struggled with this problem until Martin-Löf's breakthrough.

Nies, A. *Computability and Randomness*. New York: Oxford University Press, 2009.

> This is a readable introduction to the field.

Schnorr, C. P. *Zufälligkeit und Wahrscheinlechkeit: Eine algorithmische Begründung der Wahrscheinlichkeitstheorie*. Berlin: Springer, 1971.

> This is a very important contribution, focusing on computable martingales and proving connections with other notions of randomness. There

is an argument for a concept of randomness slightly more permissive than that of Martin-Löf (1966). Non-German speakers will have to get this from secondary sources, for instance, Bienvenu, Shafer, and Shen (2009) and Nies (2009).

Solomonoff, R. J. "A Formal Theory of Inductive Inference Parts I and II," *Information and Control* 7 (1964): 1–22, 224–54.
  Computational complexity is invented in the service of inductive logic. This contribution was mostly ignored until Kolmogorov found out about it and gave it its due credit.

CHAPTER 9

Bell, J. S. "On the Einstein–Podolsky–Rosen Paradox," *Physics* 1 (1964): 195–200. Reprinted in J. S. Bell. *Speakable and Unspeakable in Quantum Mechanics*. Cambridge and New York: Cambridge University Press, 2004: 14–21.
  Bell's nonlocality theorem for the Einstein-Podolsky-Rosen experiment.

Ehrenfest, P., and T. Ehrenfest-Afanassjewa. *The Conceptual Foundations of the Statistical Approach in Mechanics*. Ithaca: Cornell University Press, 1959. Translation of "Begriffliche Grundlagen der statistischen Auffassung in der Mechanik," in *Encyklopädie der mathematischen Wissenschaften*, Volume IV/2/II/6. Leipzig: B. G. Teubner, 1912.
  This review article on statistical mechanics was supposed to be written by Boltzmann. After his suicide, Felix Klein asked the Ehrenfests to take over the project. It is very clear and well worth reading today.

Einstein, A., B. Podolsky, and N. Rosen. "Can Quantum-Mechanical Description of Physical Reality Be Considered Complete?" *Physical Review* 47 (1935): 777–80.
  A famous thought experiment that has now been mathematically analyzed by John Bell and experimentally realized. After this, read Bell (1964).

Ornstein, D., and B. Weiss. "Statistical Properties of Chaotic Systems," *Bulletin of the American Mathematical Society* 24 (1991): 11–116.
> Dynamics and deterministic chaos by leaders in the field.

CHAPTER 10

Goodman, N. *Fact, Fiction and Forecast*. Cambridge, MA: Harvard University Press, 1955.
> An influential restatement of the Hume's problem of induction by the twentieth-century philosopher Nelson Goodman. This contains the famous *grue-bleen* example—thought by some to be a paradox.

Hume, D. *An Enquiry Concerning Human Understanding*, Section IV, Part I. London: Millar, 1748/1777: 21.
> The clearest statement of his inductive skepticism by the greatest English-speaking philosopher. A pleasure to read.

Jeffrey, R. "Probable Knowledge," in I. Lakatos, ed., *The Problem of Inductive Logic*. Amsterdam: North Holland, 1968.
> Richard Jeffrey introduces his philosophy of radical probabilism and his "probability kinematics" model of belief change with uncertain evidence.

Savage, L. J. "Implications of Personal Probability for Induction," *Journal of Philosophy* 64 (1967): 593–607.
> The author of *Foundations of Statistics* speaks to philosophers on Hume's problem of induction.

IMAGE CREDITS

Chapter 1: Cardano: Stipple engraving by R. Cooper from the Wellcome Library, London. CC-BY-4.0 via https://wellcomeimages.org/indexplus /image/V0001004.html; 1.1: From S. Stigler, *Seven Pillars of Statistical Wisdom*, http://www.maa.org/book/export/html/116153; 1.3: Photograph by Susan Holmes; 1.5: From Persi Diaconis, Susan Holmes, and Richard Montgomery, "Dynamical Basis in the Coin Toss," *Siam Review* 49 (2007), 211–35.

Chapter 2: Ramsey: Photograph by Lettice Ramsey and reproduced courtesy of her grandson Stephen Burch.

Chapter 3: Tversky: Photograph courtesy of Barbara Tversky; Kahneman: Photograph by Audra Melton and reproduced courtesy of Daniel Kahneman.

Chapter 4: Bernoulli: From Museen an der Augenstinergasse, Basel, Switzerland.

Chapter 5: Kolmogorov: From http://images.sciencesource.com/p /12758422/AndreiKolmogorov-Soviet-mathematician-SK5416.html, © RIA Novosti / Science Source.

Chapter 6: Bayes: Portrait of an unknown nineteenth-century Presbyterian clergyman. Identified as Thomas Bayes (d. 1761) in Terence O'Donnell, *History of Life Insurance in Its Formative Years* (Chicago: American Conservation Co., 1936). Courtesy of Stephen Stigler; *Essay*:

From S. Stigler, "True Title of Bayes' Essay," preprint, May 2013; 6.5: Image available at http://xkcd.com/1132/.

Chapter 7: De Finetti: Photograph courtesy of Eugenio Regazzini; Freedman: Photograph by George M. Bergman and reproduced courtesy of the Archives of the Mathematisches Forschungsinstitut Oberwolfach.

Chapter 8: Martin-Lof: From the Archives of the Mathematisches Forschungsinstitut Oberwolfach; Solomonoff: Photograph courtesy of Jürgen Schmidhuber.

Chapter 9: Democritus: *Democritus* by Hendrick ter Brugghen, reproduced from Rijksmuseum, Netherlands. Gift of B. Asscher, Amsterdam and H. Koetser, Amsterdam.

Chapter 10: Hume: *David Hume, 1711–1776. Historian and philosopher.* Portrait by Allan Ramsay, reproduced from the Scottish National Portrait Gallery: Jeffrey: Photograph by his son Daniel Jeffrey.

INDEX